S0-BUV-379

Books by Richard B. Lyttle

PEOPLE OF THE DAWN
Early Man in the Americas

WAVES ACROSS THE PAST
Adventures in Underwater Archeology

THE GAMES THEY PLAYED
Sports in History

THE GOLDEN PATH
The Lure of Gold Through History

The Golden Path

RICHARD B. LYTTLE

The Golden Path

The Lure of Gold Through History

WITH ILLUSTRATIONS BY THE AUTHOR

Atheneum 1983 New York

LIBRARY OF CONGRESS CATALOGING IN PUBLICATION DATA

Lytle, Richard B.
The golden path. The lure of gold through history

SUMMARY: Discusses the influence of gold on
history from early times to the present
and explains how gold
has been mined and crafted.
1. Gold—History—Juvenile literature.
[1. Gold—History] I. Title.
TN420.L96 1983 669'.22'09 83-6357
ISBN 0-689-31006-4

Published simultaneously in Canada by
McClelland & Stewart, Ltd.
Text set by Maryland Linotype Composition Company,
Baltimore, Maryland
Printed and bound by Fairfield Graphics,
Fairfield, Pennsylvania
Designed by Mary Ahern
First Edition

This book is for
THE POINT REYES PRINTMAKERS

CONTENTS

F O R E W O R D

No matter how well an author may know the subject ahead of time, when the writing begins there are always surprises. With this book, the biggest surprise was a golden treasure of information, too much information.

Well before the first chapter was half written, it became obvious that the story of gold is little different from the story of civilization itself. Gold has been involved in just about every human activity you can name.

Divine gold was the cornerstone for many early religious beliefs. In fact, gold worship may have begun before sun worship.

Gold has always been a favorite material for artists. Indeed its workability and its beauty by themselves have inspired many golden creations.

In written languages through the ages, gold is mentioned in everything from sublime verse to daily work ledgers. Just the legends about gold would be enough to fill a book to overflowing.

The quest for gold by dedicated alchemists was the foundation for modern science. And modern science, as if in gratitude, has proved that the early faith of the alchemists in manufactured gold was justified.

In the world of economics, gold has played the major role. The first widely-accepted money to circulate was gold. The first trade of any consequence was based on gold. The first bankers were goldsmiths and gold dealers.

In short, it became clear to me almost at once that the story of gold could never be told in one book without huge gaps in the record. I regret the gaps and beg the reader to remember that only the high points of gold's fabulous history could be described here.

In selecting these high points, I have looked for parallels between our own age and the past. You will find some of these parallels disturbing. Were Rome, Spain, and Portugal, for instance, any more foolish in their tragic dependence on gold than we are today in our dependence on oil?

We might also ask if the attempts of most modern governments to play down gold's economic role will be successful. In light of the record, it hardly seems likely. In fact, it's hard to imagine a future without the metal's golden glitter.

It seems that gold will retain its importance in the lives of men and women throughout the world for many centuries to come. That's my belief, and I think the story of gold told here will lead you to the same conclusion.

The Golden Path

A wall painting from Tutankhamun's tomb shows the young king pouring scented oil into his queen's palm as they relax in golden splendor.

Chapter One

Divine Gold

Everywhere in the palaces and temples gold sparkled and gleamed. Hammered gold sheathed furniture and statues. Gold thread shimmered in drapery, tapestry, and clothing. The very walls shone with gold.

The pharaohs and high priests of Egypt sat on golden chairs at golden tables and ate from platters of solid gold. At night they slept on gold beds. When she rose from her morning bath, Queen Hatshepsut, who ruled from 1486 to 1468 B.C., powdered her body with gold dust.

The pharaohs answered to no one except the gods. They were even thought to be divine. Early monarchs of Egypt claimed descendance from Ra himself, the sun god who rode the skies by day and battled the evils of the underworld by night. Like all important gods, Ra had several manifestations. He might appear as a hawk or a lion. Ra was always present in gold. Gold, therefore, was connected with Ra's divinity and the pharaoh's use of gold showed their divine origins.

The most enduring deity, and perhaps the most loved,

was Osiris who ruled the underworld and was the giver of knowledge and eternal life. He was entrusted with the spirits of the dead.

Osiris and his sister, Isis, according to Egyptian beliefs, were children of the sky goddess, Nut, and the earth god, Geb. Isis became the goddess of fertility and consolation in time of sorrow. She and Osiris married and had a son called Horus, who like Ra, stood for light and goodness.

The most famous myth of the Egyptian creed tells of Set, an evil son of Nut and Geb, who was driven by brotherly envy to slay Osiris, cut him into fourteen pieces, and scatter the remains throughout Egypt. Horus tracked down Set and killed him to avenge his father. Then the son became king of Egypt while the father became judge and ruler of the underworld. Egyptians, firmly believing in life after death, did everything in their power to please Osiris. And Osiris was most pleased by gold.

The common people, not blessed with gold, worshipped at the temples and gave the priests whatever they could to gain the blessings of Osiris. But without gold, eternal happiness for the commoner could not be guaranteed. The pharaohs, on the other hand, built lavish tombs and furnished them with golden treasure to be assured of the happy afterlife that was due them and to continue their link with the gods.

For the Egyptian pharaoh, all mortal life was a preparation for spiritual life after death, and one of the best ways to prepare was to amass gold and have it made into works of art that would please and impress Osiris.

Of course, gold could do many other things as well. It could control trade, raise armies, and cement treaties. It could buy both necessities and luxuries. A ruby, a bolt of cloth, or a slave could all be purchased with gold. But more important than anything else, gold could guarantee happiness after death.

A tomb mural, painted by an Egyptian artist some 3,500 years ago, shows a clerk weighing gold rings against a bull's-head counterweight.

Egyptian burial chambers were furnished with everything imaginable that the deceased might need in the spiritual world. Tables, chairs, beds, kitchen utensils, bows and arrows, spears, game boards, chariots, even boats went into the tomb. Sometimes models representing a boat, chariot, or other piece of equipment for a favorite pastime might be used, but if there were space for a full-sized boat or chariot it went into the tomb.

The tomb would also most likely contain statues of favorite gods, portrait statues of wives and children, and statues of slaves or servants to care for the deceased through eternity. Wall paintings or carvings often depicted major

This portrait of Tutankhamun, which decorated his coffin, was made with gems set in solid gold.

events or achievements in the deceased's life. Everyday happenings of household or plantation might also be pictured. Today this tomb art tells us a great deal about ancient Egypt.

The many trappings of burial such as the coffins and the jars to hold the vital organs removed by the embalmers, bore as much, if not more, gold as anything else in the tomb. There were usually three coffins, one fitting inside the other like Chinese boxes. Often each coffin bore a likeness of the deceased carved in wood and covered with gold and precious stones.

Within the inner coffin, the body, which had been treated by the embalmers for nearly three months, lay wrapped in yards and yards of linen strips. It was a common practice to place golden jewels in the wrappings. Some of the oldest golden artifacts known today were found in the linen wrappings around the mummy of a queen who died at Abydos in central Egypt in about 3200 B.C.

The sands of the Upper Nile provided an early and relatively easy source of gold for Egypt. Sifting the sands or washing them down channels lined with animal fleece were the methods usually used to gather the flakes and nuggets of river gold. But when they gave out, Egypt was forced to search for fresh deposits.

They were soon found in one of the world's bleakest regions, the Nubian Desert, stretching hot and dry from the Nile to the shores of the Red Sea. The gold, though plentiful, was bound within quartz, a hard rock that lay in formations running from the surface to several hundred feet below ground.

Tons of rock had to be mined and crushed to recover just a few ounces of gold. Obviously, it was work for slaves.

No one can guess how many thousands died in the Nubian Desert. It seems that most weak and elderly slaves never even reached the mines at all, but died during the forced march across the hot wasteland. Those who survived the march were worked to death in the very real sense of the phrase. The Egyptian overseers had discovered that it was much cheaper to bring in fresh slaves than it was to bring in food and supplies. Consequently, the work crews never received adequate food or shelter.

There were three phases of work at the Nubian mines —digging up the rock, crushing it, and washing the residue

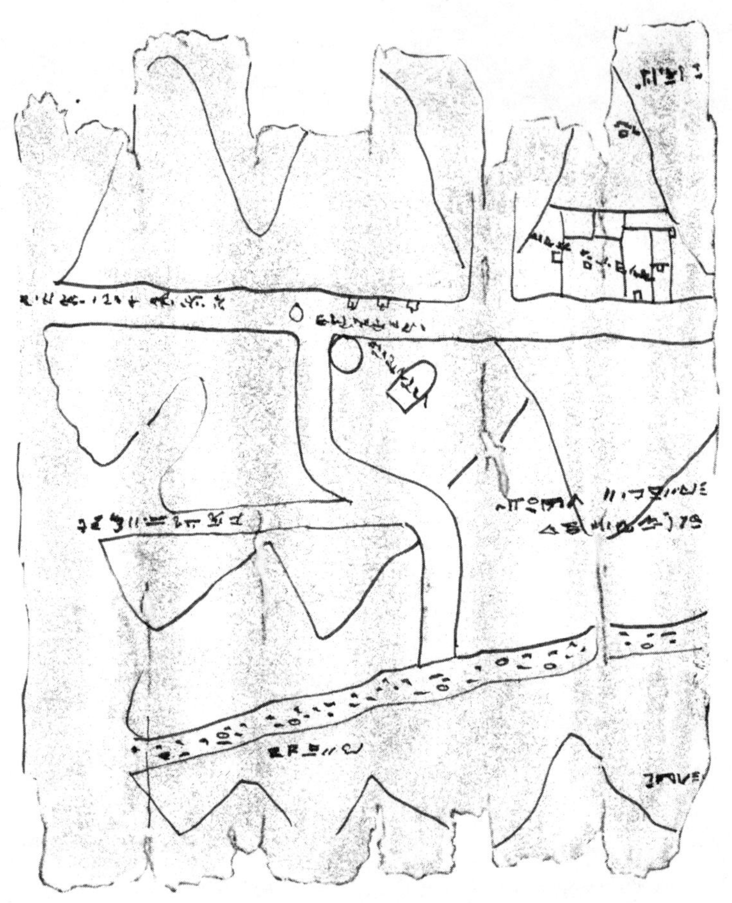

A map of a Nubian mine, cracked from centuries of rolled-up storage, shows what is probably a gold-bearing vein slanting across the lower portion.

to extract gold. The slaves sent into the mines had the most dangerous task. The shafts were nothing more than crawl spaces, just big enough for a worker lying on his back to pass up a basket full of ore.

Archeologists have found some narrow tunnels that ran for fifteen hundred feet underground, dipping as deep as three hundred feet. Temperatures rise at that depth, and with no ventilation shafts, the air would be putrid.

Oil lamps used to light the shafts used up valuable oxygen and when fires were built to crack rock, as was sometimes necessary, the oxygen supply fell dangerously low. Suffocation vied with cave-ins as the leading cause of accidental death. Sometimes, when underground fires were built to heat rock that also contained veins of arsenic, the mine would fill with deadly gas. Overseers undoubtedly cursed the time lost in removing bodies from the tunnels.

On the surface, slaves beat the rock with hammers and then passed the crushed rock to millers, who turned it under grinding wheels until it was a fine powder. The powder went to washing tables where slaves used sponges to lift away rock flakes until only gold remained. Any thirsty slave who tried to wet his lips with precious washing water was lashed without mercy.

Some one hundred separate mines have been located in recent archeological expeditions to the Nubian Desert. All were played out long ago.

But even in the days of maximum production, the Pharaohs never had enough gold.

So, using their military might, the Pharaohs could either attack neighboring nations and take their gold or force neighbors to pay tribute in gold each year as immunity from invasion. Thutmose III, who followed the gold-powdered Hatshepsut to the throne, brought the flow of foreign gold to an all-time high when he added Syria to the Egyptian Empire in the fifteenth century B.C.

Hatshepsut had sent a fleet of five large boats down the East African coast in search of the fabled "Land of Punt." After two years, the ships returned heavily ladened with gold and incense. A wall carving commissioned by the queen shows her dividing the treasure among priests and royal stewards. In addition to sacks of gold dust and gold nuggets, the horde included golden rings big enough to be worn around the upper thighs.

Punt, which may have been the gold-rich Ophir described in the Bible, has never been located by today's historians, but several African rivers apparently once contained much alluvial gold washed from rock deposits by erosion. We can assume that the pharaohs of Egypt got a good share of it.

By 3200 B.C., goldsmithing skill was well advanced. The jewels found in the queen's tomb at Abydos included four gold bracelets decorated with beads of lapis lazuli and turquoise. Two pieces bore the design of delicate flowers. Two others showed the facade of a temple topped by a perched bird.

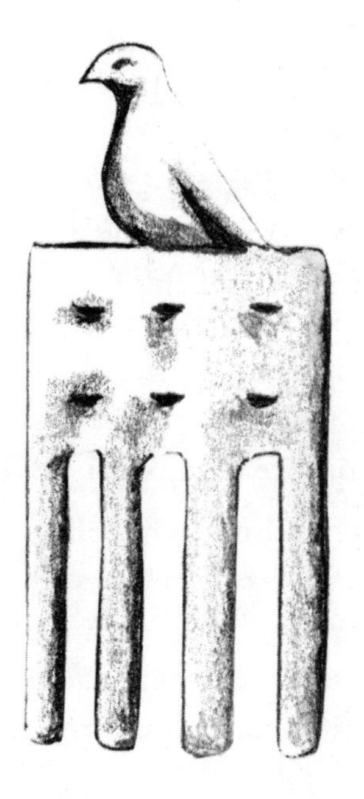

One segment of a gold bracelet made in 3200 B.C. shows a bird perched on top of a temple.

Goldsmiths probably were already being given special privileges, and it is quite likely that their skills were already family secrets, carefully guarded from one generation to the next, a tradition that survived into modern times.

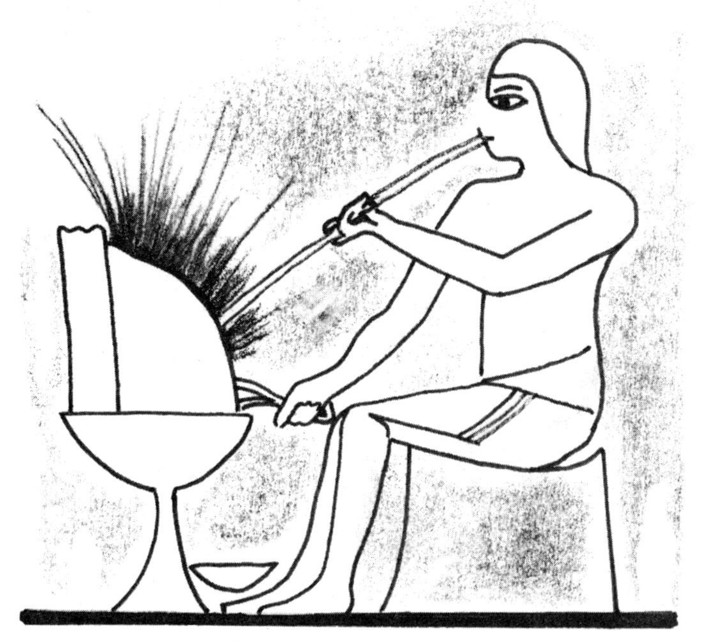

An early wall painting shows a goldsmith working at his forge. Egyptian smiths enjoyed an elite position among the ancient craftsmen.

Because it was risky to ship gold, the goldsmiths traveled to their work. Going from job to job, from city to city with news and fresh stories, the smith was much desired for his entertaining company. He was as cosmopolitan as any minstrel and received more respect because he worked with gold.

Before Egypt's long-enduring civilization began to decline, goldsmiths had mastered almost every metalworking technique known today. They had discovered that mixing copper with gold produced an alloy with a melting point almost 200 degrees Fahrenheit below the melting point of either metal in pure form. The alloy was thus ideal for soldering.

Egyptian smiths knew how to draw gold into thin threads and use the threads to create patterns of golden filagree. They could cast gold. They could beat it into paper-thin sheets. They could work sheet gold into patterns covering any solid object.

The ancient smiths were also skilled at chasing, either by engraving lines in a gold surface or by creating raised lines through pressure on the reverse side of the metal sheet. By extending the reverse pressure technique they had also invented the art of repoussé to raise three-dimensional forms from a flat surface.

Due to various impurities, gold occurs in nature in different colors. The Egyptian smiths were expert at combining different colored gold in their patterns and by 2000 B.C. they had learned to control color. Adding iron gave gold a purple hue. Copper made it red, and silver made it pale yellow. The smiths could extract silver from gold, or combine the two metals to make electrum, an alloy that gained great popularity among Egyptian royalty.

By 1900 B.C., the Egyptian craftsman could sweat thousands of tiny gold beads into glistening background patterns. This technique of granulation, as it is called, was lost for centuries, not to be rediscovered until modern times.

And by 1900 B.C., the Egyptian smiths had learned how to combine enamel with gold. They started with a pattern of powdered glass laid on a gold surface. Then they applied just enough heat to melt the glass and fuse it to the metal. Later, smiths worked long hours hammering inlays of gold into grooves etched in other metals such as bronze or silver.

Perhaps the most popular technique, however, was the painstaking art of cloisonné which called for fitting hundreds of individual, precious stones or enamel work into a honeycomb of gold cells to create intricate and brilliant patterns. The stones varied in size so that each cell of the golden framework had to be individually formed to hold it. The proudest possession of Ramses II, who ruled in the thirteenth century B.C., was said to be a golden chest orna-

Detail view of a large chest ornament depicting the god Ra shows the inlay work typical of painstaking cloisonné, which Egyptian smiths are believed to have invented.

ment with four hundred inlays of lapis lazuli, carnelian, garnet, and turquoise.

Although smithing methods and techniques improved and expanded through the centuries, designs changed very little. Like painters, sculptors, potters, and weavers, the Egyptian goldsmiths embraced tradition. This was due in large part to the long-established symbolism of the Egyptian religion. The hawk that depicted Ra, for instance, should not be changed because the god did not change. This conservative approach influenced all designs from nature. Trees and flowers, birds and beasts, were almost always depicted in the same way, with unvarying balance and proportion.

Wall carvings and paintings showed people in profile, but rarely face-on. Statues depicted people in stiff poses. Figures of women stood with their feet together and arms at their sides. Men could be shown with one foot ahead of the other, but there was little other suggestion of motion.

There was a break in this formal tradition, but it came only after Akhenaten and Nefertiti, his Asian queen, founded a new religion in about 1360 B.C. Egyptians were told to forget the old gods and worship Aten only. He stood for truth and was represented by the disk of the sun and by gold.

Truth worship encouraged naturalism in art. The pharaoh and his beautiful queen were depicted kissing, playing with their children, or conversing at the dinner table.

The new religion, which many scholars cite as the world's first venture into worship of a single god, died with Akhenaten, but artists continued to follow nature for another generation or two.

Thus the tomb of Tutankhamun, the young king who restored the old gods, was filled with a golden treasure of natural art. Even the gods were shown in natural poses. And the portraits of the young king looked as if they could speak to us across the centuries.

The 1922 discovery of Tutankhamun's tomb, practically untouched by grave robbers, brought to light the greatest collection of Egyptian art yet known. Although the style was unconventional, almost every object in the tomb was either made of or well decorated with the divine metal, the traditional stuff of royalty and the gods, the everlasting, workable, beautiful gold.

Chapter Two

Why Gold?

It was treasured long before the beginnings of recorded time. Anything rare and beautiful was likely to be desired even by primitive people, but gold has not usually been valued for rarity and beauty alone. There were other qualities.

To some, gold had magic power. It could cure illness or give knowledge. It was often thought to be divine. In some cultures gold was called the sweat of the gods. Gold stood for faith, purity, good fortune, beauty, happiness.

Several superstitions about gold have continued into modern times. If you owned a gold coin, you might well be tempted to carry it as a good luck piece. And why not? Even if it didn't bring luck, it would certainly draw the attention and perhaps envy of your friends.

But why gold?

Gold has so many unusual traits that it is hard to pick any single trait that has made the metal so important through the ages. Actually, it is probably a combination of two traits that does most to set gold apart from other metals. It is highly workable and at the same time it is almost indestructible.

No other rare metal or precious stone has this remarkable combination.

Gold can be pressed into molds, stretched into thread, pounded into sheets, and polished smooth as glass. It is not just workable, it's extremely workable.

Modern smiths describe gold as an "obedient" metal. It is as malleable as clay, but unlike clay, it will not collapse of its own weight. While copper turns spongy or snaps under heat, gold will not. While iron oxidizes rapidly and becomes hard and brittle, gold will not. While silver turns stubborn when reheated, gold can be reheated, reformed, folded, stretched, lumped, and flattened again and again.

It can be pulled into a hair thinner than the eye can see; it can be flattened into sheets no thicker than 1/300,000 of an inch.

And it mixes well with other metals. Although pure gold melts at 1945 degrees Fahrenheit, the addition of a little copper will lower the melting point, the property Egyptian smiths exploited for soldering. In addition to changing the color of gold, the addition of other metals often serves to harden it.

Egyptian smiths not only mastered many techniques, but they were also excellent artists, as this fish pendant demonstrates.

Gold will not corrode in air or rust in water. It can remain buried in the earth for centuries and be brought to light unchanged. Gold has been favored for dental work for centuries. A gold filling is good for a lifetime and then some.

The durability of gold is responsible more than anything else for the symbolism that has become attached to it. A gold wedding ring stands for lasting love. A gold watch is given at the close of a career in appreciation of fidelity. And for luck that really lasts, you want that gold coin.

Scientists, who refer to gold as Au, which comes from *Aurum*, the Latin word for the metal, can cite still more attractive qualities. Gold is heavy. It's atomic weight of 196.967 is close to lead's atomic weight of 207.19.

Gold is a good conductor of electricity and because of malleability and resistance to rust, gold is often preferred over copper and silver for television, calculator, and other fine circuitry. Transistors in space vehicles are linked with gold.

Extremely thin gold will filter damaging infrared rays from sunshine, a property now being put to use in making windows for offices and factories. Gold is first treated to make it soluble in oil. The mixture then is applied to glass, the glass is heated, and the oil burns away to leave an invisible but effective film of gold.

Because gold mixes readily with other metals such as silver, copper, tin, and iron, natural deposits of pure gold are rare. And because the addition of other metals give gold special qualities, refined gold is rarely produced in pure form.

Gold purity is usually described in karats, one karat being one twenty-fourth of the total weight of metal being rated. Pure gold is thus rated at twenty-four karats. Gold that is seventy-five percent pure is rated as eighteen-karat gold. In most jewelry shops today the gold is either fourteen-karat or ten-karat. Although degree of purity does determine purchase and resale value, it is hard for the eye to see any difference between ten- or fourteen-karat gold and pure gold.

Gold is indeed one of the rare metals, but natural deposits of it have been well distributed. It has been profitably mined on every continent of the world. Large amounts of gold are also present in seawater, but an economic method of tapping this source has not yet been invented.

There are two basic kinds of natural deposits. Thin veins of gold can be found in the seams of certain rocks. When the rocks formed under heat and pressure, molten gold was forced between the seams to be trapped there when the rock cooled. Quartz is the most common gold-bearing rock. Therefore the mining of such deposits is often described as quartz or hard rock mining. Usually the layers of quartz gold are so thin that they cannot be seen with the naked eye. The rock must be crushed and processed to recover the gold.

Alluvial gold is the product of erosion. It has been washed out of its original place in rock formations and deposited as dust or nuggets in river sands or gravels. Recovery of alluvial gold can be done by pan, rocker, sluice, or dredge. These methods rely on the pull of gravity that causes heavy objects to settle faster in water than lighter objects. A prospector panning for gold has only to agitate the mixture of sand and water in a whirling motion that gradually spills the lighter sand and gravel from the pan. Eventually, only heavy gold remains at the bottom.

Alluvial mining, obviously easier than the digging and

crushing of quartz mining, was undoubtedly the method used by the primitive people who first began gathering gold. We don't know who these people were or exactly why they gathered gold. Through the ages, from one culture to the next, attitudes toward gold have varied tremendously.

New World natives, for instance, did not equate gold with wealth at anything near the same degree the Conquistadors did. The Aztecs and the Incas liked the stuff for its color and the easy way that it could be worked into statues, jewelry, and other ornaments. Thus, when Cortez and Pizarro demanded gold, the natives complied readily.

This leads us to the irony of gold. Despite all its fine qualities, gold has more often than not brought out the worst in people. Men and women have committed all kinds of crimes for it. They have killed for it. They have risked their lives for it. Many have died for it. Slaves have been doomed for it. People have fought for it. Many have married for it while many others have left home for it.

Great civilizations have often been ruined by gold. Nations rich in natural deposits of gold spent unwisely. Luxury items were imported at inflation prices. Rising costs of labor and materials undermined industry. The economy survived only as long as more gold could be produced. And more gold meant higher prices.

When the deposits of gold gave out, as they always did sooner or later, the nation was left without a solid base for its economy. With no industry, no goods to export, and no source of well-trained labor, the nation sank into poverty.

Spain, as we shall see, is a classic example of the harmful effects of gold. The country was fabulously rich in domestic deposits and was the first to reap the golden harvest of the New World. But Spain today is one of the poorest countries of Europe.

Generally, nations that have not been blessed with gold fared better in the world marketplace. The only way they could obtain gold was by trading with countries that had it, and trade backed by a healthy industry gave a solid foundation for economic growth. England is the classic example of a nation that benefited from lack of natural deposits of gold. The nation built an empire on trade, and in the process it filled the vaults of its London banks with quantities of gold.

In the ancient world, we can look to one of the earliest civilizations for another good example of an empire built on trade, trade that rose out of hunger for gold.

Chapter Three

Golden Towers

Given so little to work with, it is remarkable that the Sumerians, builders of the first true cities, achieved so much.

Their one gift from nature was the rich farmland in the valley formed by the Tigris and Euphrates Rivers. The region, known as Mesopotamia, lies today within the boundaries of Iraq.

Although floods laid down new deposits of nutritious silt, the rivers were not predictable as was the Nile. Long periods of drought could wither crops and bring disaster to the ancient farmers.

Yet, the civilization of Mesopotamia thrived. The people were both energetic and inventive. They created a written language and used it widely for everything from record-keeping to epic poetry. They wrote laws, made fundamental discoveries in astronomy, and established several rules of numbers that still serve mathematicians today. The Sumerians also invented the arch and the wheeled vehicle.

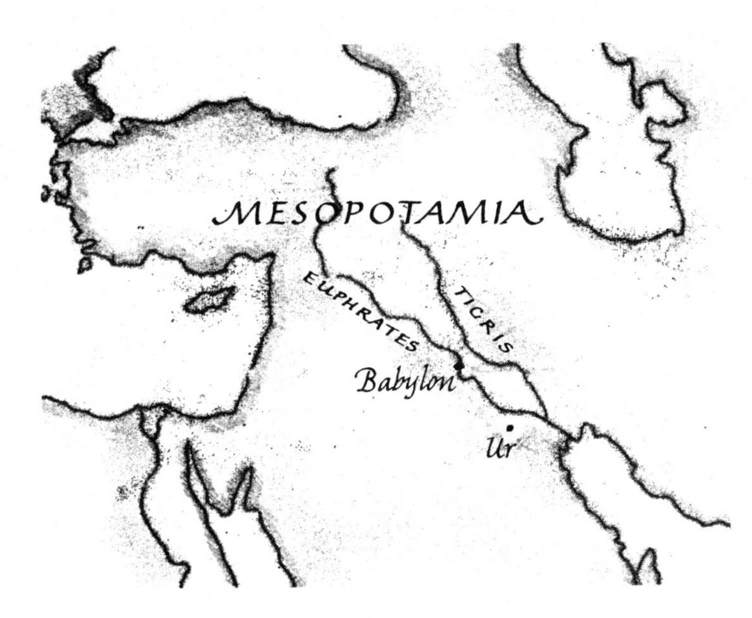

The fertile floodplains of the Tigris and Euphrates nourished the Sumerians and other early civilizations, who established trade to obtain gold.

By today's standards, however, the Sumerians were a deprived people. Stone and wood were so scarce that they had to build with mud bricks baked in the sun. Their territory had very little copper and lead and no silver, no gold. Yet the early records tell us that the Sumerians valued gold highly. And they got it, heaps of it.

The civilizations arose from a mixed stock of hunters and herdsmen who over the centuries had wandered down from the north, discovered the valley, and became permanent residents. It may well have been the mixed ethnic heritage that gave the people their vitality and inventiveness. By about 3000 B.C., the fundamentals of civilization were clearly evident.

Farmers had long before banded together to dig communal irrigation ditches and help one another with harvests and livestock roundups. Cooperative action not only increased yields but also produced a new and strange commodity—spare time. The Sumerians turned spare time to good purpose. In addition to a written language and other inventions, they refined the crafts. Potters made exquisite vessels. Tanners produced leather of the finest quality. Dyers gave wool bright new colors. And weavers made textiles unlike any the world had yet known.

To see the work of Sumerian craftsmen was to desire it, and this situation, of course, prompted trade. Sumerians exported leather, raw wool, cloth, pottery, and in good harvest years, grain. Trade brought all the things the Sumerians lacked—wood, incense, silk, copper, lead, iron, silver, and gold.

The caravan routes extended east to India north into Armenia, west to the shores of the Mediterranean, and south to meet Arab seamen who traded along the east coast of Africa. The donkey caravans also went regularly into Egypt.

Trade brought the bonus of learning and craftsmanship from other lands. Sumerians welcomed and were stimulated by foreign ideas.

Thus, as gold arrived in the Euphrates so did the knowledge of how it could be worked. Smithing skills developed by the Egyptians were quickly adopted and perfected by Sumerians. But it was not entirely a one-way connection. Recent archeological evidence suggests that the Sumerian smiths may have invented methods that helped the Egyptian gold workers expand their craft.

We don't know, for instance, if the Egyptians or the Sumerians were the first to use the refining process known as cupellation. It called for mixing the gold to be purified with rock salt, lead, bran of barley, and sometimes tin. The mixture, placed in mud-lined pots, was baked five days.

Under prolonged heat, the salt fused the gold into a single mass. The bran removed all organic impurities that would burn while the lead formed a slag of nonburnable impurities. The tin, when used, served to harden the gold. When the vessels were removed from the ovens, their covers were lifted and the slag was skimmed away. The molten gold that remained could then be poured into molds for casting.

It is hard to believe that this complex method could have been developed independently in Egypt and Sumeria.

The Sumerian smiths soon learned that the purity of the gold depended on degree of heat in refining. And they may have invented the control of color by adding small amounts of other metals.

Sumerian smiths joined gold pieces with either solder or rivets. Often their large pieces were made out of several cleverly joined castings. They used open molds for flat castings and closed molds for three-dimensional shapes. They also mastered the lost-wax process.

First step for a lost-wax casting was to make a figure of wax. This was then set in soft, molding clay and fitted with a pouring vent. After the clay hardened, molten gold was poured into the vent. The heat of the gold vaporized the wax, leaving a void to be filled with metal.

When the casting cooled, the clay was broken away, leaving a golden replica of the original wax figure. Probably the most remarkable thing about lost-wax casting is that the technique seems to have been invented independently in several regions of the world. Native Americans were using it long before Columbus.

Sumerians, who hammered out large pieces of sheet gold, learned that heating the metal kept it from turning brittle under the hammers. They also found that two large sheets could be joined by pounding, a process known today as annealing.

Another method that was refined, if not invented, by the Sumerians relied on a form of bitumen as a foundation for a three-dimensional image in gold. After thin sheets of gold were formed over the bitumen, it was melted away, leaving the golden image as a hollow shell.

Usually a Sumerian smith worked with many other metals beside gold. Their bronze was the best in the ancient world. At first they produced it by smelting ores that contained both copper and tin or copper and arsenic, but as early as 3000 B.C., the Sumerian smiths were producing top quality bronze by combining tin and copper in fixed proportions. The copper came from present-day Turkey, the tin all the way from the Caucasus Mountains in Russia.

The Sumerian civilization, like that of Egypt, was guided by strong religious beliefs. Devotion to the gods is particularly evident in Sumerian crafts. Religious symbols dominated the designs. Craftsmen and artists were convinced that any skills they had were given only to glorify the gods.

The civilization was run by a powerful priesthood. Only priests, it was believed, could converse with the gods. Any order from a priest was thus divine and had to be obeyed.

Priests owned most of the irrigated land and directed use of labor. Slaves toiled endlessly, building temples to the gods. A group of temples formed the center of the cities and the temples themselves were almost always dominated by a tower or ziggurat. These did not look like Egyptian pyramids nor were they built to house the dead. The typical ziggurat was a series of stacked square or circular structures each smaller than the one below so that they rose to a peak high above the surrounding buildings and walls. Each level was joined by steps or a ramp.

Every unit of the structure was a room of worship. The top room, being closest to heaven, was the most divine and thus the most richly decorated.

Made of baked brick, ziggurats usually had drab exteriors, but the doors into each room were decorated lavishly either with bronze or sheet gold, and the interiors gleamed with rich furnishing.

Marble facing, sheets of gold, or tapestry covered the interior walls, and thick carpets covered the floors. Often each room in a ziggurat held the statue or statues of principal gods.

Although we have no verbal descriptions of a Sumerian temple, we do have accounts of ziggurats built by the Babylonians who later ruled in Mesopotamia. The Tower of Babel described in the Bible was one of several ziggurats built by the Babylonians. Although it once towered almost three hundred feet high, Babel's tower is now just a heap of rubble some seventy miles south of present-day Bagdad.

The ziggurats were just one of several temples built by the Babylonians. When their capital city of Babylon was at its peak, it had 53 separate temples, 955 shrines, and 372 altars. The main temple, 470 feet long, dated back to 1800 B.C. It was called *Esagilia* or "house of the lofty head," and it contained giant golden statues of the gods Sapanat and Marduk.

The temples were repeatedly raided by invaders and repeatedly refurbished. The Hittites carried off the statues, but the Babylonians recovered them. Later, the Assyrians often galloped off with temple gold.

Heroditus, the Greek traveler and historian, visited Babylon in the fifth century B.C. He reported that the city stretched on both sides of the Euphrates. The capital was circled by a wall so thick that a four-horse chariot could be driven on top of it. The wall had many turrets where bowmen could shower attackers with arrows. There were some one hundred gates of bronze leading into the city.

Although Persian armies had left much of Babylon in ruins prior to Heroditus's description, he evidently talked to many people who recalled the splendor of the temples. The main temple, the historian wrote, was plain on the outside except for the gold and bronze doors that opened into each chamber.

Inside, the walls were covered up with marble below and black, red, and blue murals above. Golden canopies hung from the ceilings, and the golden statues were shrouded in cloth woven from gold thread. The statue of Marduk, whom Heroditus called Zeus, was portrayed as a giant seated on a golden throne beside a table and footstool made of solid gold. Heroditus estimated that the gold furnishings inside the temple weighed eight hundred talents which translates to forty-eight thousand pounds. He also said that another statue of Marduk outside the temple was made of solid gold and stood twenty feet high. Exaggerations?

Heroditus probably was a little casual with the truth on some occasions. In fact "the father of history" was sometimes more commonly known as "the father of lies." But there can be no doubt that the ancient people of Mesopotamia did value gold and knew how to get it and work it, particularly for religious purposes.

Archeological proof of the golden wealth of the Sumerians was discovered early in this century. The find, made in the ruins of Ur at the south edge of the Mesopotamian plain, did much to vindicate Heroditus.

For years, the many mounds that dotted Iraq were not recognized as ruins. Even after treasure hunters learned that the mounds were often the best places to hunt for ancient relics, travelers did not realize that the low mounds on the horizon marked the remains of cities built of baked brick.

England's Sir Leonard Woolley, one of the first to lead a scientific dig in Iraq, chose an isolated cluster of broad mounds that appeared totally lacking in importance. Although some of the mounds were sixty feet high, even relic hunters had ignored them.

The unimposing mounds, it turned out, marked the site of Ur, an ancient city that had been occupied for two thousand five hundred years. And the mounds hid an archeological treasure in gold.

Sir Leonard worked at Ur from 1922 to 1937. Test holes drilled into the mounds during his first season revealed that in about 4300 B.C. Ur's earliest inhabitants were living in mud-daubed huts. This earliest layer of habitation lay beneath some eight feet of silt that revealed no evidence of man.

Clearly, Ur had been drowned beneath a huge flood, a flood that may well have covered most of Iraq and been the very flood described both in early Sumerian writings and in the Bible.

When people returned to the city, their culture was apparently similar to that of the first inhabitants. The new arrivals lived in simple huts built of mud-daubed sticks. But as the centuries passed and levels of occupation built up one upon the other, the new above the old, stone tools improved, building became more ambitious. The first temple, a low ziggurat, was built about 3000 B.C.

Fascinating discoveries were made daily at Ur. Sir Leonard had the valuable assistance of his wife, but no other trained staff. They both worked long hours to identify and date each find. When there was time they tried to piece together the story of a civilization that lasted far longer than anyone had dreamed.

In their first season, the Woolleys located several tombs, but their excavation had to be delayed because other projects demanded full attention. In 1927, however, the diggers went to work on the tombs.

The crew of one hundred forty native workmen excavated an area eighty yards long and sixty yards wide. It proved to be a well-populated cemetery. A total of eighteen hundred separate burials were eventually identified. Some of the deepest graves dated back to 2800 B.C. Most had been damaged by robbers, but a few were undisturbed, and vases, tools and other funeral relics proved that the people of Ur had a strong faith in the afterlife.

The richest and most shocking discoveries came from the royal graves. There were fifteen of them, and they not only contained remains of a king or queen but also the bodies of servants and guards who went to the grave with their masters, apparently voluntarily.

Like the other burials, most royal tombs had been looted, but one had never been disturbed. It was the tomb of Pu-Abi, a Sumerian queen who died about 1700 B.C. Her husband, King Abargi, lay nearby, but his grave had been cleaned out by robbers.

Queen Pu-Abi was surrounded by golden treasure. With her were the bodies of sixty-three servants, soldiers, and musicians. They lay in a pit at the end of a long ramp. A crypt had been provided for the queen's body, but the others lay on reed mats that covered the floor and walls.

After months of excavation and careful study of each artifact, the Woolleys were able to reconstruct Pu-Abi's burial ceremony.

As musicians played, a sledge bearing the queen's

A small cup still contained traces of eye makeup when the Woolleys found it in the queen's tomb.

body was hauled down the ramp into the pit. There the body was lifted from the sledge and placed in the crypt. Soldiers, musicians, waiting women, and maids then filed into the pit. Next came chariots pulled by donkeys and attended by drivers and grooms.

When all were in position, each person drank from a cup that contained either a deadly poison or a narcotic. When workers began filling the pit with earth, the servants were either already dead or blissfully sedated against the horror of living burial.

The Woolleys found the bodies of nine women lying against the Queen's crypt. Apparently ladies in waiting, they each wore elaborate headdresses made of gold, carnelian, and lapis lazuli.

Nearby, ten women lay in two even rows and beyond them was a harpist. She wore a golden crown and her harp was beautifully formed and lavishly decorated with gold. The remains of a chariot lay near the entrance to the pit with the bodies of two crushed donkeys and their grooms.

The body of Queen Pu-Abi lay among golden treasure.

There was a fourteen-inch dagger of solid gold with a hilt of lapis lazuli inlays. The golden sheaf bore a filagree design. There were vases, cups, and bowls, all of gold. One small bowl contained a green paste that probably was used as eye shadow. Other personal items included a golden head scratcher, a pair of tweezers, and a small golden spoon for cleaning the ears. There were countless beads, many golden jewels, and fine samples of repoussé and granulation.

The most remarkable single piece, however, was the queen's delicate headdress. When reassembled and placed on a model, the headdress seemed to float like a towering halo. A row of golden rings had graced the dead queen's brow. Above dangled three rows of leaves made of thin gold with finely etched veins. Above the leaves was a row of flowers made of white calcite and blue lapis lazuli, and topping them was a row of gold and lapis lazuli flowers. Carnelian chains linked the flowers with the leaves. The queen had also worn golden earrings in the shape of large crescents that complimented the headdress perfectly.

The Woolleys concluded that the headdress was too delicate to have been worn in life and that it was made especially as a burial ornament.

Although the headdress was an outstanding example of goldsmithing skill, Sir Leonard prized another artifact even more. In a nearby tomb, where a king named Meskalam-dug had been buried, was a helmet that had been hammered from a single nugget of fourteen-karat gold. It was designed to fit low on the head and had cheek pieces to protect the face. To suggest hair, the ancient smith had etched lines and welded strands of golden wire on the helmet surface.

Where did the Sumerians get their gold? Analysis of the finds at Ur suggest that most of it came from Egypt as

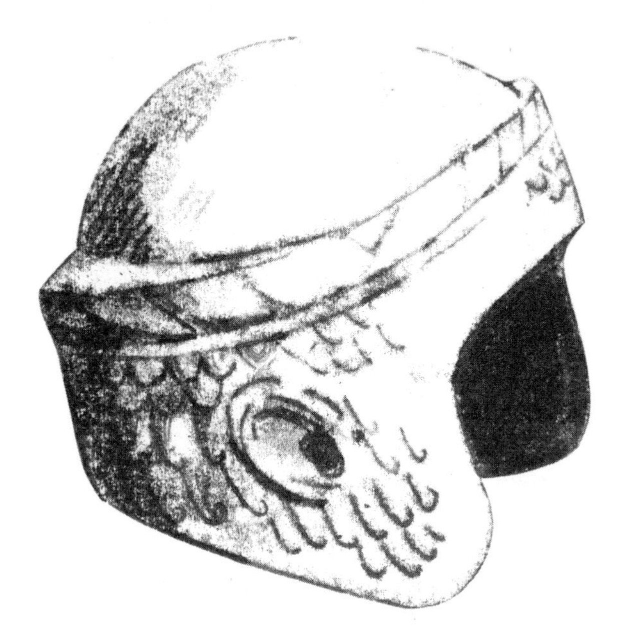

Sir Leonard's favorite artifact was a helmet hammered from a single gold nugget.

payment for trade goods. But the Sumerians also obtained gold from three mining areas in Arabia. One area was near the Gulf of Aqaba, another far south near the mouth of the Red Sea, and the third to the east not far from the Persian Gulf. Again, the Sumerians probably obtained gold from these regions through trade.

Like the towering ziggurats of the region, the civilizations of Mesopotamia were built one on top of the last. The Babylonians, the Assyrians, the Medes, and the Persians all followed the pattern of empire set by the Sumerians. All used trade to gain gold, and the gold amassed by one civilization fell into the hands of the next until there was a huge treasure locked within the storehouses of the Middle East.

Westerners long coveted the golden horde, but no one was able to touch it until the fourth century B.C. when a boy king rode out of Greece to avenge his people. His name was Alexander, and we will get to him in Chapter Five.

Chapter Four

Legendary Gold

When we call a prospector an "argonaut," speak of the "Midas touch," or describe someone as being "rich as Croesus" we are reaching back to golden tales of ancient time.

Although the Greeks were the first to record these tales, they describe events that took place in or near Asia Minor.

Jason, sailing in the *Argo,* led his "argonauts," gold-seekers, through many adventures and narrow escapes until he reached Colchis, a kingdom on the east shore of the Black Sea in what is now part of Georgian Russia. There, Jason and his men stole the famous Golden Fleece or ram's skin and returned to Greece.

The gods had promised Jason that an ancient kingdom would be restored to him and his heirs if he brought the fleece home, but that did not happen. Driven from home by a despotic ruler, Jason wandered for many years before his strange and final rest.

The legend says that the hero, weary of travel, lay

down beside the hulk of his rotting *Argo*. While Jason slept, the heavy bow timber fell and crushed him to death.

Although the much-told story of Jason is largely myth, it contains foundations of truth.

For one thing, the skins of sheep or other furry animals were used to line sluice boxes where river sands and gravels were washed for their gold. Flakes of gold that lodged in the wool were usually recovered by drying and then burning the fleece.

The legend of Jason also reflects the lack of gold on mainland Greece. To find gold, the Greek prospectors had to embark on voyages to distant lands. Indeed, it was hunger for gold as much as anything else that led the Greeks to trade, colonize and eventually dominate the Mediterranean world. Undoubtedly the story of Jason, told before the home fires, gave more than one Greek youth the restless urge to travel.

King Midas was thought for years to be nothing but a character of fiction, but recent studies reveal that there really was a King Midas—actually several of them. The Midas dynasty ruled over the land of Phrygia, located in what is now central Turkey.

The people had come from the Balkans and settled the land in about 1200 B.C. Phrygia was a poor country until gold was discovered in the Pactolus River.

How did the gold get there? According to the legend, the first Midas to rule the land was like everyone else, a very poor man. But then he befriended Silenus, a stranger who turned out to be the foster father of the god Dionysus (Bacchus to the Romans). To repay the king's kindness, the god offered Midas one wish. The king, who had long dreamed of wealth, asked that everything he might touch would turn to gold.

Dionysus granted the "Midas Touch," but it brought quick disaster. When the king tried to eat food, it turned to solid gold before it reached his mouth. Facing the certainty of starvation, Midas next took his beloved daughter in his arms and watched in horror as she too turned to gold. The grieving king begged to be freed of his magic power. Dionysus told the king to wash himself in the Pactolus river. Of course, as soon as Midas entered the river, its sands turned to gold and the king's awful magic ebbed away.

The gold of the Pactolus made Phrygia and its people rich. They dominated the region from the Eighth Century B.C. until well into the Sixth Century, when they were at last brought under the rule of their neighbors to the north, the Lydians.

Lydia, in what is now northwest Turkey, was also rich in alluvial deposits of gold, and most of it apparently was amassed by Lydian royalty. It was Croesus, the last king, who became the symbol of personal wealth. To be "rich as Croesus" was to be well fixed and then some. Lydia, however, was famous for more than the legendary wealth of its rulers. It is believed to be the first country to standardize gold coinage.

Actually, the origins of coinage go back long before the Lydian kingdom. For centuries, the most common medium of exchange was livestock. Five cows might buy a roll of woven wool or forty sacks of flour. We don't know when symbols for cows first replaced the actual animals. But obviously a live cow was difficult to take about on a shopping trip, and it was impossible to divide.

Notes of exchange, perhaps scratched on a piece of wood or stamped in wet clay, were most likely the first form of money. If Merchant A gave Merchant B a note promising to deliver five cows on the first day of the following month to pay for forty sacks of flour and Merchant B then gave the note to Merchant C to buy a roll of woven wool, it is fair to say that the use of money had begun.

But it wasn't standardized. The notes said nothing about the quality of the cows, the purity of the flour, or the design of the woven cloth involved in the three-way transaction. And the merchants in our example would not be likely to make such a transaction with strangers. There had to be trust.

The problem was solved somewhat by the use of generally accepted tokens for commerce. In the ancient world, a common token was the cowrie shell. Most cowries came from the shores of the Indian Ocean, far enough away for the shells to retain some scarcity. Of course, the shells were easy to carry and store, impossible to counterfeit, and readily accepted as a medium of trade by strangers as well as neighbors.

We don't know when or where gold and other precious metals began being used as a medium of exchange, but we can assume that the practice had crude beginnings. An unstamped ingot of copper, perhaps even a lump of unrefined iron ore were among the earliest forms of metal money.

The first refinements called for molding or stamping the metal into recognized shapes and sizes. Gold rings of known weight were favored by Egyptians. Such rings were usually molded by the smelters at the mines.

Later, gold and silver shaped like cowrie shells became a common medium of exchange. Because of its rarity, beauty, and endurance, however, gold was always preferred over other metals. The softness of the metal contributed to its popularity.

By chopping off a portion of a gold bar, payment on any transaction, no matter how small, could be met. A thin slice from a bar, of course, produced a disk. The Babylonians standardized the weight of such disks to produce the famous gold shekel. It weighed about 8.34 grams. Palestine soon had its own shekel, weighing more than sixteen ounces.

Blank disks of gold, however, presented problems.

In the hands of a sharp dealer, a blank shekel could be shaved of some of its gold, and over the years a coin shaver could eventually amass a fortune. With no telltale marks, the shaved coins continued in circulation at less than their original value. Merchants, to avoid being cheated, began weighing coins as a regular practice with each transaction.

In addition to weight variations, the purity of gold also varied. While one sample might contain fifty percent gold and fifty percent silver, another might be just thirty percent gold and seventy percent copper. While color of the two samples would be different, it took a highly trained eye to detect the difference. Thus, because of variations in purity, gold remained suspect until the touch test was developed.

The testing equipment consisted of three sets of twenty-four gold needles, each of known degree of purity. One set was made up of needles of gold and silver, another of gold and copper, and the third of gold and various mixtures of copper and silver. Usually arranged in a ring, the first needle of each set contained one part gold and twenty-three parts of the baser metal. The amount of gold increased with each needle around the ring until the last which contained pure or twenty-four parts gold. We need look no further than the touch needles to find the source of the karat measure for gold.

To test gold with the touch needles, the sample was first rubbed against a dark stone with just enough pressure to leave a yellow mark. Then the tester used the set of needles to make a mark that matched the color left by the sample. The needle that matched the color closest not only showed the percentage of gold in the sample, but also told the tester if the baser metal was copper, silver, or a combination of both.

Although simple and accurate, the touch test could not be done quickly. And weighing gold again and again also cut into a busy merchant's time. To save time, merchants adopted a system of symbols that could be stamped on the gold stating its weight and its purity. Lydian merchants of the eighth century B.C. are believed to be the first to use such symbols on privately issued coins.

In the next century, King Allyates ordered the first public issue of Lydian coins. Government backing, symbolized by the likeness of a lion's head on the cast coins,

The lion's head symbolized government authority and thus appeared on many early coins.

increased trust, and consequently broadened circulation of the coins. Marks on the coins, particularly if stamped on both sides, also discouraged the practice of shaving.

Even the government coins, however, were not standardized. Made from electrum with a high proportion of silver, the gold content of the coins might vary from thirty-five to fifty-five percent. Although the purity was marked on each coin, quality variations from one coin to the next made trade difficult. The situation was further confused by the continued issue of private coins which also lacked standardization.

It was not until the middle of the sixth century B.C. that Croesus took the confusion and risk out of coinage by establishing a bimetal system. He separated the gold and silver from the native electrum and minted coins of standard purity in each metal.

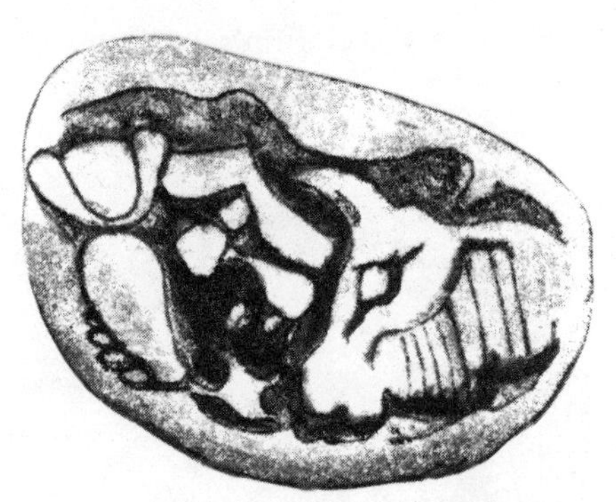

The Croesus stater bore the likeness of a lion and a bull.

The Croesus stater, oval-shaped and bearing the heads of a lion and a bull, were the world's first international coins. They soon inspired other countries, particularly Greece and Persia, to adopt official coinage of their own.

Persia was to take much more than coinage from Lydia. In 546 B.C., the Persian army invaded the little country and annexed it to the Persian empire. Thus ended the reign of a legendary king, but the legacy of Croesus remains today, jingling in our pockets.

A twenty stater coin, shown actual size, was issued in Bacteria in Western Asia in 160 B.C. It may be the largest gold coin ever made.

Chapter Five

Persians and Greeks

Reverence for gold as a divine metal changed gradually into desire for gold as wealth and money. Most of this change in attitude occurred during the turbulent centuries of Greek and Persian rivalry.

Historically, the gold-rich Persians and other early peoples of the Near East and the gold-poor Greeks are difficult to separate. Two cultures, though vastly different, were deeply intertwined in art, thought, commerce, and events.

The first historical event, recorded by Homer perhaps three centuries after it occurred, was, of course, the Trojan war. Legend says that the Greeks crossed the Aegean to attack Troy in order to rescue Helen who had been stolen away by Paris, the handsome Trojan prince. But the conflict most probably rose from a long-standing trade dispute.

The Trojans, who controlled the entrance into the Black Sea, where some shores glittered with gold, had either charged heavy duty on Greek shipping or stopped passage of Greek boats entirely. This, historians believe, was the true reason why the Greeks launched their attack.

Up until the middle of the last century, however, most historians discounted Homer's *Iliad* as a complete myth. Not only did the war never happen, but Troy, scholars said, never existed. But one man, the son of a German preacher, believed Homer had described real events.

Heinrich Schliemann's driving goal was to discover Troy. Although born of a poor family in 1822, he had a brilliant mind and an excellent business sense. When he saw that his first job in a grocery store from 5 A.M. to 11 P.M. each day would lead nowhere, he ran away from home, worked in Amsterdam, and then sailed for America.

He was shipwrecked and almost died of exposure before returning to Amsterdam to take an excellent job with a dye company. His boss soon picked him to represent the firm in St. Petersburg, Russia. There, Schliemann set up his own business and began studying languages. He made almost a million dollars and mastered six languages before he sailed again for America, headed this time for the California gold fields.

In just nine months as a banker in Sacramento, where he dealt in gold dust and nuggets, he made another fortune. Then he returned to Europe to trade in tea, cotton, and other commodities and amassed even more wealth.

At age forty-one, after learning nine more languages, he decided he was ready to begin the search for Troy. Convinced that the Trojans had hidden gold in the walls of their besieged city, Schliemann, with a crew of one hundred fifty Turkish diggers, began work in 1871.

The German merchant's belief that Troy lay beneath a mound near Hissarlik on the Aegean coast proved correct. But in four seasons of digging, the workers uncovered not one ancient city but nine different levels of occupation.

One morning in June of 1873, while digging in the

By the time he was forty-one, Heinrich Schliemann had amassed a fortune and was ready to begin his quest for ancient Troy.

ruins of the second oldest city, Schliemann uncovered a fabulous treasure that had indeed been buried beneath the ruins. In all, there were 8,763 pieces of golden jewelry, including earrings, necklaces, rings, and headdresses. One elaborate headdress was composed of 16,353 delicately formed pieces of gold.

Schliemann and his young wife, Sophia, a native of Greece, smuggled the treasure to Athens and thereby launched a long, legal dispute with Turkish authorities. Meanwhile, the discovery brought worldwide attention with both acclaim and criticism.

Scholars, who had once claimed that Troy never existed, now damned the discoverer for destroying so much of the ruins in his quest for gold. And indeed, Schliemann had dug through and destroyed much of the layered record to reach the ruins of what he believed was Homer's Troy.

Actually, it has since been proven that the second oldest level in the mound dated back to about 2500 B.C.,

A golden beaker, bent by the weight of ruins, was one of hundreds of gold artifacts Schliemann discovered.

long before the Trojan war. Just the same, Schliemann's work at Troy and later at Mycean ruins in Greece woke up many drowsy scholars and laid the foundations for modern archeology.

The Trojans were just one of the peoples beyond the Aegean who rivaled the Greeks. Although today we often refer to the rivals collectively as Persians, many civilizations and many people rose and fell from power in the Middle East, and the Persian Empire itself was composed of many different people.

The once proud Phoenician cities along the eastern coast of the Mediterranean and the Israel of Solomon had fallen to the cruel and aggressive Assyrians. By 900 B.C., the Assyrians had extended their boundaries to include Turkey, Syria, and the lower regions of the Mesopotamia. By 700 B.C., the Assyrian kings ruled from the banks of the Nile to the Persian Gulf. Slaves, tapestries, silver, and gold flowed into the Assyrian city of Ninevah, capital of the

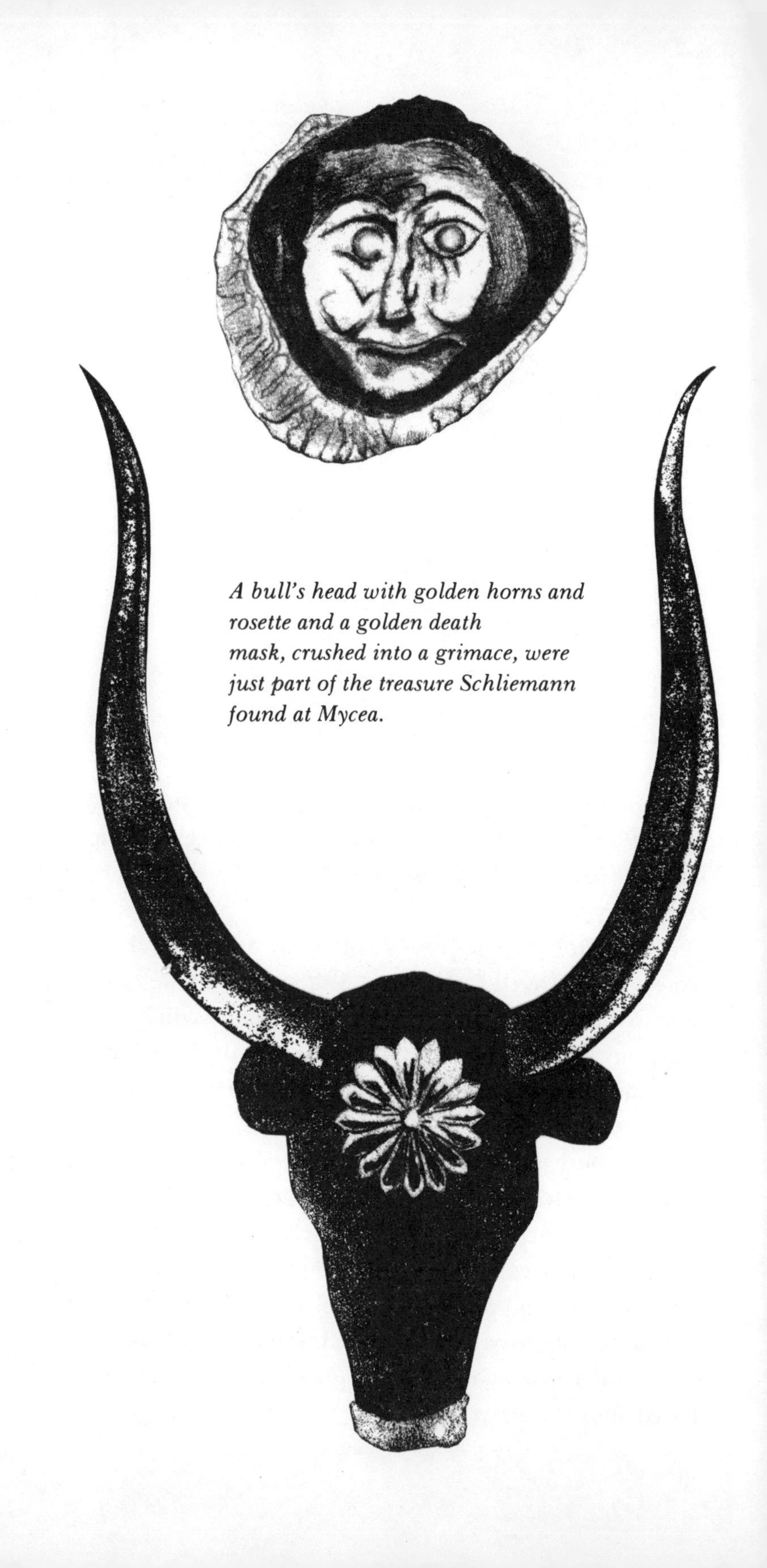

A bull's head with golden horns and rosette and a golden death mask, crushed into a grimace, were just part of the treasure Schliemann found at Mycea.

largest empire the world had yet seen. Under threat of invasion, neighboring states were forced to send tribute in gold.

Peoples who did not submit to Assyrian rule were slaughtered. The brutal diplomacy did not win the Assyrians many friends, and when internal conflicts weakened the empire, the neighboring Medes, a tribe that had migrated from India centuries earlier, sparked a revolt on the northern boundary. Then early in the sixth century B.C., the Medes united with the Babylonians to lay siege to Ninevah.

The city, surrounded by a moat and double walls eight miles in circumference, reportedly held out for three years, but finally, Sardanapalus, the last Assyrian king, realized his cause had no hope. He gave his five children three thousand talents (about 171 thousand pounds) of gold and ordered them to flee. Then he arranged a legendary suicide.

The king's slaves began stacking wood. The work continued for several weeks until the pyre was four hundred feet high with a level top some one hundred feet long. The workers then built a chamber of incense cedar on top of the pyre.

Sardanapalus placed the accumulated treasure of his empire inside the chamber. Countless gems, gilded furniture, golden bowls, plates, and utensils, gold jewelry, and ornaments representing the art styles of all regions of the empire, were toted up to the chamber. By some accounts the furnishings included one hundred fifty golden couches and one hundred fifty golden tables.

After purple tapestries woven with gold were hung to complete the arrangements, the king and all members of his harem climbed a wooden stairway to the chamber. Then the huge pyre was ignited.

The blaze, which darkened the sky with smoke for fifteen days, appeared to the people of Ninevah as a sacrificial fire within the palace walls. They learned only after the city had fallen that their king had killed himself.

Although the story may be pure fiction, we do know that Assyria fell abruptly, and the Medes and the Babylonians divided the spoils and territory of empire. The Babylonians, who controlled lower Mesopotamia, were in cultural decline. Soon the Medes united with the Persians, another tribe of Indian background, to overrun Babylon.

The Persians, under the leadership of Cyrus, rose to dominance and eventually extended the new empire from the eastern shores of the Mediterranean into northern India. Their boundaries included King Midas's Phrygia and Croesus's Lydia. Even after the Persians gained control of all gold-bearing regions of the known world, except those in Spain and a few mines owned by the Greeks, the expansion policies continued. Cyrus died fighting in central Asia in 529 B.C., but Persian aggression never slackened. Darius, who came to power in 521 B.C., was the first to issue gold coins with a royal portrait. He became known as the "King of Kings" throughout and beyond the empire.

Persian coinage, which always held its value, helped hold the vast empire together. Other factors contributing to unity were fair and efficient administration, and a provincial system of government. Each province or satrap, as it was called, was ruled by a governor directly answerable to the king. The king relied on generals and spies to make sure that his governors did not abuse their trust.

The system worked well through many generations. It proved particularly efficient for raising gold. Each province was responsible for so much tribute each year. The king's stewards had only to weigh the gold as it arrived at the counting house.

Meanwhile, the Greeks expanded their influence through trade and colonization, first around the shores and islands of the Aegean and then westward into the Mediterranean. Some of these colonies, particularly those on Sicily, produced gold, much more gold than came from the mines on Thrace and Thasos in northern Greece. But compared with Persia, Greece was always gold poor.

Silver from mines near Athens was relatively abundant, and the Greeks founded their monetary system of silver. What gold could be had was reserved to furnish the temples.

Unlike the early Greeks, the Persians did not worship gold. Instead they coveted it for its power in trade and diplomacy. Although Persian smiths were highly skilled in making ornate ornaments and jewels, most of the raw gold brought in by the caravans was melted down and poured into ingots for storage. It could then be converted to coins as needed. And sooner or later it was needed. The Persian hunger for gold could never be satisfied.

Under Darius, the Persians attacked Greece and captured Macedonia in the north and the gold mines of Thrace, but further advance was blocked, first by storms and later by the gallant stand of a handful of Greek heroes at Marathon in 490 B.C.

Ten years later, however, Xerxes led the greatest invasion force ever assembled toward Greece. Lavishly financed by the provinces—Lydia alone contributed seven thousand pounds of gold and ten thousand pounds of silver to the campaign—the army bridged the Hellespont and marched into Greece. A vast fleet sailed down the coast, keeping pace with the army. A suicidal stand by Spartans at Thermopylae failed to stop the Persians. Even the loss of most of his ships at Salamis failed to turn Xerxes back.

He entered Athens, plundered the temples of their gold and set fire to the city. For seven months his army looted and slaughtered wherever it went. Xerxes left Greece a defeated land, but the Greek people had been roused. They hungered for revenge.

More than a century and a half passed, however, before the Greeks were able to take their revenge. And the instrument of revenge, surprisingly, came from Macedonia, that northern region, which had always been considered wild and barbaric. Alexander the Great was raised as a prince of this "backward" realm.

His father, Philip II, had already united most of Greece and impressed his enemies with his skill at arms. The Macedonians had a new combination that no other army could match. For defense, Philip's army used the phalanx, a unit of closely packed spearmen who presented a wall of overlapping shields. For offense, he used light, swift cavalry.

When Philip died in 336 B.C., Alexander was only twenty, but he lost no time in putting his power to use. After snuffing several minor revolts with swift and decisive campaigns, he crossed the Hellespont with thirty thousand infantry and five thousand cavalry. The Persians waited with a force of one million. The year was 334 B.C. Alexander never returned.

Under Darius III the Persian defenders were not well organized. The Greeks won a series of small battles as they marched toward Syria. Several coastal towns had fallen to the invaders before they met the main Persian force near the present Syrian border.

The six-foot-six Persian king, robed in splendor on a

A gold coin issued during the conquest may bear the truest likeness of the young hero, Alexander.

golden chariot, was so confident of victory that he had brought his queen and his mother to witness the battle. He had also brought baggage trains of gold. His huge army, representing all regions of the empire, surged forward like an angry sea.

But if the Persian horde was the sea, Alexander's well-disciplined band was the rock. The Persians struck and scattered. The Macedonian cavalry charged and routed Darius's loosely organized army. The king himself fled, leaving his women and most of his gold. The Greeks won an undreamed of treasure. They also won entry to the entire Middle East.

True, some cities resisted the Macedonians, but Alexander had already set a pattern of fair treatment and wise diplomacy. He respected the Persian gods and temples. He encouraged growth and prosperity. Often he left the existing political leaders in power, trusting them to follow his orders.

Many cities opened their doors to the Macedonians and received better treatment than Persian rule had provided. A few cities, like Tyre, resisted, earned Alexander's wrath, and were sacked. When Tyre fell all the men were slaughtered and the women and children were sold into slavery.

Although he did treat most conquered cities with a soft hand, he never failed to take a firm hold on municipal treasuries. Most of the major cities had huge treasuries.

Sidon, Tyre's neighbor on the Mediterranean coast, reportedly yielded so much gold that Alexander was forced to bury it. The pack train he planned to send back for the treasure never arrived. If the story is true, a vast store of gold still waits discovery today somewhere near the ruins of the ancient town.

The Persian goldsmith who made this lion drinking cup used two separate pieces of gold. Cup and lion were then joined with an almost invisible seam.

The greatest treasure of gold was taken from Persepolis, the Persian capital. Alexander himself reported that it required five thousand camels and twenty thousand mules to haul the loot. The Persian riches included gems, robes, tapestry, furniture of wood or ivory overlaid with gold, gold jewels and ornaments, gold coins, and gold bars.

To revenge Xerxes' burning of Athens, the Macedonians set fire to Persepolis.

Alexander, handsome, young, athletic, and full of restless energy, continued eastward, winning victory sometimes by diplomacy and sometimes by force of arms. By supporting existing religious beliefs, often even worshipping the local gods, he gained allegiance from the conquered people.

If diplomacy called for it, he never hesitated to marry a local princess. To the distress of his generals, Alexander also began wearing Persian clothes and adopting Persian customs.

He led his men across the Indus River into India and would have gone farther, perhaps into China, but his weary army refused to continue. Alexander pleaded but was soon forced to return to Babylon.

There he held court for his few remaining years. He encouraged many of his men to marry Persian women. He received diplomats from throughout the realm, he interviewed scholars, he encouraged trade and the arts, and he drank wine. It was probably inactivity and boredom that led to Alexander's excessive drinking. At more than one party he would compete with his generals to see who could down the greatest quantity of wine.

In June of 323 B.C., weakened by drink, Alexander fell sick, probably with influenza. Within a few days he died. He was thirty-one years old.

The huge empire he had united was soon divided into

three parts by his generals and their families. Egypt went to the Ptolomies, Syria and the east went to the Selucids, and Macedonia and the rest of mainland Greece went to the family of Antigonus. Despite division and frequent border battles, there was a flowering of trade, prosperity, and culture as the Greek and Persian worlds joined.

Egypt's Alexandria, with a library established by the Ptolomies, became the new center of learning. Scholars from all regions were welcome and they gathered there by the hundreds.

Greek scholars, stimulated by eastern thought, made great advances, particularly in mathematics and astronomy. Art and literature were also stimulated to new heights.

Greek became the international language and Greek literature was deeply influenced by eastern writings. Scholars at Alexandria, fired by restless energy, translated the records of many different people into Greek. The laws and history of the Jews, translated by the Greeks, became the foundation for the teachings of St. Paul. The Greek translation of what is now called the Old Testament was Christianity's first Bible.

Thus it can be said that Alexander's conquest and unification of eastern civilizations gave the West cultural and spiritual treasures that far outweighed all the gold that he liberated from Persian vaults.

Inevitably, the flowering of culture eventually faded and the ruling families of Alexander's legacy became decadent and corrupt. By the time Roman legions marched eastward, the houses of the Ptolomies, Selucids, and Antigonus were ready to topple.

Chapter Six

Roman Gold

In the eighth century B.C., the small settlement among the hills overlooking Italy's Tiber River certainly did not look like the seat of empire. Poverty, more than all else, set Rome apart from the neighboring Etruscan villages.

But the leaders of the settlement were ambitious. They organized the community under strict laws, called themselves kings, and slowly began to bring the neighboring farm communities under their influence. Rome's expansion policy, necessitated by a fast-growing population with an ever-growing need for new land, had begun.

By the sixth century B.C., with the power of Etruscan and other tribes waning, Rome set its goal on a unified Italy. Tribe after tribe fell in defeat or simply submitted to Roman dominance. After each new territory was taken and looted, it was repopulated by Roman farmers.

While land motivated the early expansion, gold soon rose in importance. Though gold had been scarce under the Etruscans, the Romans learned to treasure it not for spiritual reasons, and not particularly for its beauty, but for its value. Gold represented wealth and power. It could buy land, goods, and slaves. There was never enough of it.

By 450 B.C., the Romans had decided that gold was too valuable to be used in burials. Laws banned the ancient practice. The laws, however, did little to solve the gold shortage.

It was not until Rome extended its influence into northern Italy that a steady supply of gold became available. The alluvial deposits there, long worked by the local tribes, began yielding treasure to the Romans. But the Romans had discovered an easier way to obtain gold. Their method was little more than robbery and blackmail. They called it conquest.

The method was simple. When the legions annexed a new territory, temples, public buildings, and treasuries were stripped of gold. The poor were sold into slavery. And to avoid the same fate, the rich paid for Roman protection with gold.

When the legions moved south into the boot of Italy, the Greek colonists who had long been settled there were allowed to remain only as long as they paid tribute in gold. By then, Rome was more interested in gold than new land.

When Rome annexed the Balkans, north of Thrace in northern Greece, it was to gain control of deposits that had long supplied the Greeks with gold. And when Rome began eyeing the lands bordering the western Mediterranean, it focused particularly on gold-rich Spain.

Rome's western ambitions, however, were thwarted by Carthage, the empire that had grown from a Phoenician colony on the North African coast. With a strong maritime heritage, Carthage had become a sea power. Its fleet commanded respect and fear.

Rome was a nation of farmers. Its people were rooted

to the earth. Its armies fought on land. But Roman ambition recognized no barriers, not even the alien sea.

At first, Carthage discounted Rome. How could a people without a navy be a threat? It was a joke. Hoping to avoid senseless bloodshed, Carthage sent a peace mission to Rome. The ambassadors were amazed. Not only did the Romans lack a navy, but they were also desperately poor. In their official house calls, the visitors saw just one silver service, and it was being shared among several families.

Back in Carthage the ambassadors reported that Rome posed no threat to empire.

But the Romans began building a fleet to challenge Carthage. The campaign began in 264 B.C. Early success gave the Roman sailors confidence, but then bad luck and inexperience combined to bring disaster. After the loss of sixteen hundred ships, the few remnants of the fleet scurried for shelter. Rome, with its treasury empty, was left without defense.

In a last and desperate gamble, members of the Roman senate donated gold from their own savings to build a new fleet. Although Roman sailors went into battle with more hope than promise, they had gained experience. When they met the enemy fleet they drove it from the sea. Carthage sued for peace, ending the first of the three Punic Wars.

Rome's victory brought fabulous spoils.

Carthage gave up Corsica, Sardinia, and Sicily, which became the first Roman provinces. The island people were forced to send gold tribute and grain to Rome. Carthage itself paid some fifty-seven million dollars in gold, but more importantly, Carthage allowed Rome access to some gold-bearing regions of Spain. The mines of Spain were to supply Rome for the rest of its gold-hungry days.

The mining in Spain drastically increased the need for

slaves. At first glance, it might seem that Roman slave masters were more humane than Egyptians. And in the beginning, when slaves were scarce, this was true. But with increased conquest and greater and greater numbers falling under the Roman yoke, slaves became expendable. Many were worked to death, friendless in a foreign land.

Roman campaigns often were nothing more than slave roundups. Generals grew rich as the entire population of a conquered territory was sold to the slave masters. The wise general spread much of the new wealth among his legionnaires to assure their allegiance, and some common soldiers received so much gold that they were loath to return to their farms. This led to a gradual change in Roman economy, a change that saw small, independent landholdings being swallowed up by large estates. With both land and gold falling into the grasp of a select few, social divisions became more sharply defined.

The second Punic War broke out in 218 B.C., and even though Hannibal and his elephants marched into Italy for a long occupation, Rome eventually won the major battles. A defeated Carthage paid another huge penalty in gold and gave Rome full control of Spain.

Wealth poured into Italy. The rich, overwhelmed by the bounty, lost all reservation in the display of gold and jewelry. Roman ladies and gentlemen decked themselves in golden rings, armbands, and heavy necklaces as they vied to outspend their neighbors. Banquets, another form of competition, called for exotic dishes such as bird tongue or camel heel stew flavored with foreign spices. Of course, all was served on plates of gold. Overindulgence was the test of a successful banquet. Guests ate and drank until they collapsed.

Although homes of the rich were decorated with

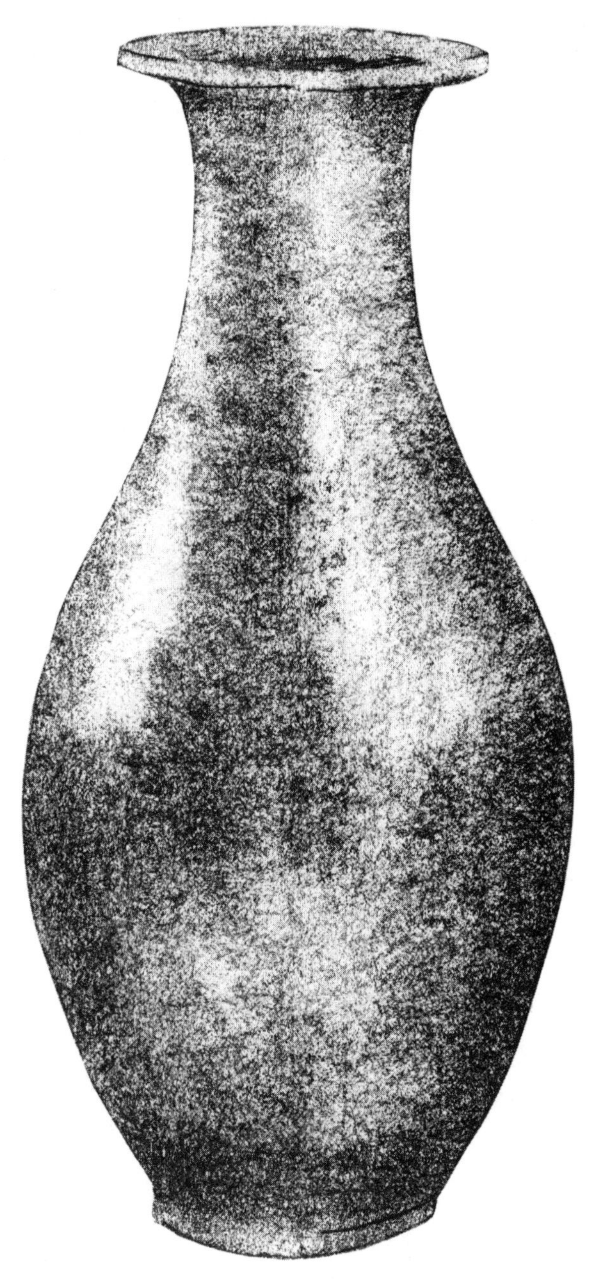

Although most Roman gold work was too gross for good taste, the vase, made in the first century A.D., *was an exception. Known as the Harwood vase, it was recently sold in London for $27,700.*

imported tapestries and silks, the Roman aristocracy did not leave a legacy of good taste. Their best designs were borrowed from the Persians or the Greeks, but usually the urge to make a lavish display of gold or precious stones destroyed the delicacy and balance of such designs.

The luxuries, of course, had to be purchased with gold, and because Rome had few craftsmen, most luxuries were imported. Trade never approached true balance and this was a serious weakness of empire. Rome's major industry was conquest. When it ceased, Rome's fate was inevitable. Remarkably, few Roman leaders recognized the problem until it was too late. During all the years of conquest, success made them blind.

In 194 B.C., leading an army financed with Spanish gold, Flaminius brought Greece into the empire. The campaign gained countless thousands of slaves and a treasury in gold.

The campaign also gave Roman aristocracy a passion for Greek art—not just because it was beautiful, but because it was rare and therefore represented wealth. Roman carvers learned to copy Greek statues while Roman goldsmiths adopted Greek designs and methods.

One Greek design, calling for open or pierced patterns of gold, lent itself well to the Roman love of large medallions. Black enamel fused into lineal patterns etched in gold, another Greek technique, also became popular in Roman jewelry.

Roman artisans perfected another Greek technique called fire gilding which made it possible to cover large walls and pillars with a layer of gold. The process was costly in human lives. Gold powder was first mixed with mercury and then painted on the surface to be gilded. A fire torch held against the surface vaporized the mercury and left the gold behind, fused in a gleaming, paper-thin coating.

Workers, who could not avoid breathing the vaporized mercury, died a slow death. The first symptom of mercury poisoning was loss of teeth and hair. Decay of nervous system, loss of mental powers, and death followed. Today, all civilized countries have banned mercury gilding for humanitarian reasons.

Rome had no such laws. The emperor Domitian, who ruled in the first century A.D., spent the equivalent of 22 million dollars to gild the roof and doors of the temple of Jupiter on Rome's Capitoline Hill. The cost in human lives from this venture was not recorded.

Nearly all Roman gold in or on public buildings and the huge private collections of gold jewelry and utensils were lost when the empire fell to barbarian hordes. Today, the best samples of Roman goldsmithing art are found in the coins.

The first Roman coins were minted by generals in the field to pay the armies. It was a simple thing to melt down looted gold, pour it into blanks and stamp the blanks with a die. Persians and Greeks had earlier minted coins for trade, but the Romans were the first to make coinage an exclusive function of their generals. The practice raised serious political problems.

Although a few gold coins were issued as an emergency measure during the second Punic War, the Roman general Sulla was said to be the first to mint large quantities of gold. His record illustrates both the good and evil facets of the practice. He showed that minting was simple. The tools were easily carried on a campaign. The army could be paid at once and in full. The risks of shipping gold long distance from Rome were avoided.

The evils, however, outweighed the advantages. Sulla's gold coins, issued in the first century B.C. after he defeated Mithridates VI in Asia Minor, set the pattern of self-glorification that was to govern coin design for centuries. Sulla's first coin showed the goddess Victory placing a crown on his head. No politician could ask for better advertising.

When Sulla returned to Rome, he brought home fifteen thousand pounds of gold and one hundred fifteen thousand pounds of silver taken from Greece and Asia Minor, and he was followed by a fiercely loyal faction. He quickly bought his way to power and ruled Rome from 82 to 79 B.C. It was a bloody era. Sulla, with a team of assassins on his payroll, killed any who opposed him. He ordered the death of some two thousand, seven hundred, including ninety Roman senators. The assassins were paid up to ten thousand dollars for each job, but usually the loot confiscated from the victim more than covered the cost.

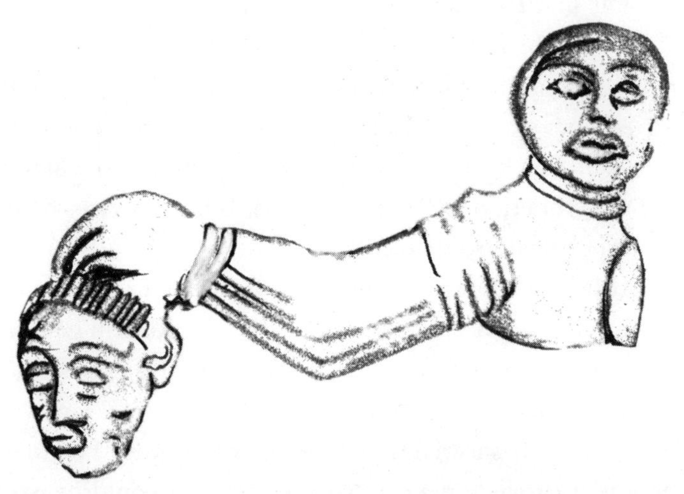

A gold-plated pin, used to hold a toga at the shoulder, was probably worn as a badge by some Roman soldier who wanted to remind enemies of their fate.

The power to mint coins, put in the hands of politically ambitious generals, was devastating to the empire. The senate tried to curb the influence of the generals, but for some reason the government was reluctant to take over exclusive control of coinage.

As long as it was necessary to buy votes, running for office was extremely expensive. Many a Roman went into debt to finance campaigns. If they lost, their creditors promptly took over their property. Crassus, one of the creditors who rose to power on the backs of political failures, financed the career of young Julius Caesar.

After conquering Gaul, Caesar returned to become one of Rome's best rulers. During his reign the empire was extended to England, where the legions found little of the rumored gold. Rich mines in Ireland remained beyond the Roman grasp.

The murder of Julius Caesar in 44 B.C. plunged Rome into civil strife that continued until Augustus Caesar rose to power in 27 B.C. Augustus made coinage the official function of government, taking the long-abused power out of the generals' control. Both silver and gold coins, issued in great number, became the standard of exchange for all business throughout the empire. Augustus, greatly helped by the influx of Egyptian gold during his reign, put Rome on solid financial foundations and restored prosperity and confidence.

Augustus was also one of the few rulers to set an example of restraint in personal spending. Compared with emperors before and after him, he lived simply. And he was careful about accepting gifts. When the cities of the empire sent him thirty-five thousand pounds in gold for his coronation, he saw to it that every ounce was returned. There was tribute enough without special gifts.

The imperial coins of Rome were modeled on the pattern earlier established by the generals. One side would show a portrait of the emperor while the other would illustrate one of his achievements. Even people in the most distant provinces knew the face of their ruler.

Eventually there were some six hundred state mints, each housed in a temple dedicated to Juno Moneta, a name that provided the root for "monetary" and "money." Although counterfeiters were hanged or fed to the lions when caught, there were always some false coins in circulation. The practice of shaving the edge of coins, however, was effectively discouraged by the introduction of the milled edge, a Roman invention that remains in use today.

Despite the sound financial practices of Augustus, the Empire continued to spend gold as fast as ever. There was never enough to satisfy the demand, and the demand fell most heavily on the gold regions of Spain.

Until about the second century B.C. the gold of Spain's rivers could be recovered with incredible ease. All it took was a sharp eye for a glittering nugget or the gleam of a thick layer of gold dust. One "mined" gold simply by bending over and picking it up.

Along the Guidalquivir River, children sometimes brought in more gold than their parents. Even after most visible gold was recovered, washing the sands produced high yields for many years. The sand was run through sluice boxes lined with brush. At the end of the day, the brush was burned and the gold recovered. Eventually, however, recovery became more difficult.

By the time of Augustus, when alluvial deposits started to give out, the Romans were digging for gold. Hard rock mining called for refined engineering skills and the large gangs of slaves. Shafts had to be dug deeper and deeper to reach the rich veins. Pumps had to be designed to draw off troublesome groundwater. Some mines plunged six hundred fifty feet below the surface.

Heat and poor ventilation took a horrible toll of slaves. In most cases the workers were locked or sometimes chained in the shafts so they could not flee at night. Their only escape was death. And many did die in the mines, where their bones are still being discovered by today's archeologists.

The Roman mine at Aramo, with a vertical shaft as the only entrance, has the richest deposit of bones. When work there ended for the day, it seems that the slave masters simply removed the ladders and went home. In the morning, those slaves who had survived the night were put back to work. The dead lay where they had fallen.

Local populations in the mining regions were soon depleted, forcing the Romans to import slaves from North Africa. The bewildered arrivals soon found themselves crawling through cramped tunnels to take their place in basket passing chains. When the baskets of ore reached the base of a shaft, they were cranked to the surface by windlass.

The Romans used mercury to recover gold from ore. The mixture of crushed rock and mercury was dumped on leather sieves. The gold united with the mercury and seeped through the sieves, leaving the unwanted rock behind. Like the Sumerians and the Egyptians, the Romans refined their gold by cooking it in ovens with salt and sulfur—the old process of cupellation.

But the Romans exploited a new and devastating process for mining gold. It is known today as hydraulic mining, and in most countries today, laws prohibit hydraulicking of any kind, but in Roman Spain there were no restrictions.

Resourceful Roman engineers dammed streams and piped water to the minefields. There, forced by gravity through leather hoses and jets at great pressure, the water carved away the topsoil. Gold-bearing veins of rock and buried gold-bearing sands were quickly exposed. The method, far less costly in its toll on slaves, was soon in use throughout Spanish mining regions.

Whole mountains were carved away. Streams were clogged with mud. On the coasts, natural harbors filled with silt washed down from the mines. Many of these harbors, once vital to the economy, could never be used again. Their loss was to contribute to centuries of Spanish poverty.

Pliny, the Roman naturalist and writer who visited Spain in the first century A.D., reported that the mines were producing fourteen thousand pounds of gold a year. He did not cite figures on the damage that was being done, but a recent study estimates that fifty million tons of Spanish topsoil was washed away by Roman hydraulics.

Despite the costly production of gold, there was never enough to satisfy Rome. All the mines in the Balkans, Arabia, Egypt, Gaul, and Spain could not meet the Roman demand.

The example of thrift set by Augustus was forgotten soon after his reign ended. With Nero, who ruled from 54 to 68 A.D., Roman aristocracy entered an era of spending more frenzied than ever before.

When prospecting expeditions he sent to Ethiopia, the shores of the Black Sea, and far into Asia all returned without gold, Nero fell back on Sulla's methods. It became dangerous to live in Rome if you were rich and out of favor with the emperor. Nero even ordered the deaths of some of his friends in order to confiscate their estates and pay his debts.

When he began running out of wealthy victims, Nero decided to save gold by using less of it in Roman coins. The devaluation of money repeated by later rulers put the Empire on the fateful road to inflation.

The government might have mended its economy with thrift and the encouragement of local industry. Instead, a great flow of gold continued to pour out of Rome to buy both the luxuries and the necessities provided by distant traders. The provinces grew rich while Rome sank deeper into debt.

All hope of reform was dashed early in the second century A.D. when the conquest of what is now Romania and Hungary brought a new influx of riches. The area, known as Dacia to the Romans, yielded some five million pounds of gold and ten million pounds of silver in loot alone. Roman control of the Dacian mines would prove a far richer treasure.

The Dacian conquest, however, was the last to line the coffers of empire. When the third century dawned, Rome once again was in desperate need for fresh sources of gold. None could be found.

The provinces, which had benefited from Roman spending, began to feel their strength. A weakened Rome

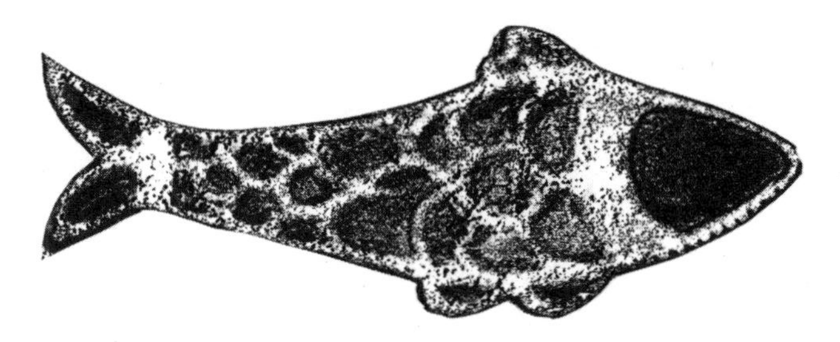

A fish pendant, made in France during the sixth century A.D., illustrates the decline of art as well as empire. It is a far cry from the lifelike Egyptian fish shown on page 17.

could not control its distant territories. Soon Rome could not even marshal adequate defense against the restless pressure of northern barbarians.

Actually, it is remarkable that Rome survived into the sixth century. The last to rule, Romulus, closed the book of empire in 576 A.D. when he retired to his farm after selling the throne to a Germanic general for a healthy sum of gold.

Chapter Seven

Alchemy

In keeping with its mysterious nature, the origins of alchemy are shrouded in the past. We do know, however, that both Egypt and China produced some of the earliest records of the art. It seems that soon after the desire for gold arose, people began trying to make it.

By some accounts Hermes Trismegistus, a mythical Egyptian priest, was the first to write about alchemy. One venture he was said to have described was his discovery of the Hermetic seal, which he used to close two halves of an egg that was later hatched into gold.

The priest was identified with the Egyptian god Thoth or Tehuti, who stood for divine wisdom. He was also loosely connected with the Greek god Hermes, the messenger of Olympus, who stood for invention as well as wisdom. Hermes became Mercury to the Romans.

Hermes Trismegistus gave his name to the "Hermetic art," an alternate term for alchemy, and "Hermetic gold," the product of the art. The word alchemy itself was probably derived from the old Egyptian *khem*, used to describe the fertile black land bordering the Nile, where almost anything could be made to grow. The Arabic article *al* was later added to the word. The root *khem,* of course, also gave us our word *chemistry.*

Whether or not they invented alchemy, the Egyptian practitioners certainly made a great contribution to the lore and traditions of the art. Unfortunately, the writings of the Egyptian alchemists did not survive. They may have been lost when invaders burned all the books of the library at Alexandria in the seventh century A.D. Another story, however, says that the Roman Emperor Diocletian destroyed all the writings on alchemy to punish the Egyptians after their unsuccessful revolt in the third century A.D. Loss of the great library, in any case, closed a long chapter in scholarship.

As a seat of all learning, Alexandria attracted serious alchemists as well as mathematicians, astronomers, and philosophers. It was at Alexandria that the two schools of the Hermetic art arose. Greek scholars, traditionally searching for life's meaning and guiding principles, encouraged the theoretical school. The Egyptians, however, were more interested in practical matters.

While the Greek alchemists explored theory, the Egyptian alchemists experimented with metals, dyes, and other useful compounds of trade and industry. The practical alchemists often operated combination dye and smithy shops. Such shops were among the earliest working laboratories of science.

When an Egyptian alchemist changed through chemical reaction the color of a compound he demonstrated the Greek's theory of transmutation. All agreed that it was through transmutation that Hermetic gold would one day be produced from less valuable elements.

Color change could be achieved in many ways. Malachite, the green copper carbonate, could be changed into lustrous copper. Galena, the common gray ore of lead, often contained substantial quantities of silver and zinc. When alchemists refined such ore in their shops, the zinc and silver they produced from lead ore was further proof of transmutation.

When heated, the red ore cinnabar changes dramatically into quicksilver or mercury, the liquid metal that combines readily with most other metals and has a particularly strong attachment for gold. Alchemists were fascinated by mercury. They thought that studying its behavior would reveal all the secrets of transmutation.

Although they were on the wrong track, the practical alchemists invented many techniques and developed much

The alchemists invented many vessels, heaters, vats, and other strange gear for their experiments. Some inventions are used by today's chemists.

laboratory equipment that has come down to modern science. Stills, ovens, beakers, and flasks were just as vital to alchemists as they are to today's chemists.

Alexandria's Maria Prophetessa, an early woman alchemist, won fame for her invention, a water bath for compounds that became known as a bainmarie. Today's chemists distill their compounds with a bath little different from the bainmarie.

In China, the earliest mention of alchemy dates back to the second century B.C., when the Emperor Lui Ch'e began searching for divine gold. The emperor wanted this

Chinese emperors often believed that eating off golden plates would give immortality. As far as we know, this did not work.

special gold, which evidently came from heaven rather than the lowly earth, in order to gain immortality. He thought that eating off plates made of divine gold would make him live forever.

The alchemist Li Shao-chun may have been responsible for this fiction. Certainly he did nothing to discourage it. In fact, he told the emperor that a beautiful woman dressed in red had given him the secret of divine gold.

The emperor provided a laboratory and all the materials Li Shao-chun requested, but there the account ends. We only know that Lui Ch'e never achieved immortality, and that Chinese alchemy remained more spiritual than practical.

The Chinese believed that gold could cure illness as well as give immortality. Gold's intrinsic value, according to the Chinese alchemists, was secondary to its healing power. But there were skeptics. The Chinese word for alchemy was *kem-mai,* which also means to wander or go astray.

The most famous of the Chinese alchemists was Ko Hung, who wrote a book late in the fourth century A.D. in which he described his discoveries, telling among other things how to produce divine gold. He also told of many gold alloys and was the first to describe a sulphide of tin known as mosaic gold.

Ko Hung seems unique. Most of his fellow alchemists were so bogged down in mysticism that they were not taken seriously by other Chinese scholars.

About the time of Ko Hung, changes in alchemy were taking place in the western world. The changes came with the shift of power and influence from Alexandria to Constantinople, capital of the Byzantine Empire. Unfortunately, one change was to shroud knowledge in secrecy.

Alchemy became a closed fraternity. Strange symbols, including pictures of fierce dragons, war chariots, red salamanders, and green lions were explained only to a select few. A green lion, we have since learned, stood for Aqua Fortis, an acid that dissolved gold.

Although Greek scholars, with their clear and open logic, continued to teach alchemy in the Byzantine world, much of the writing about the art was in code that no layman could decipher.

All this changed under Arab influence. As the forces of Islam spread across North Africa and into Spain, Bagdad emerged as the new seat of learning. There alchemy was described in plain language with no mystery. Although this was due partly to the Islamic belief that symbols were evil, the openness can be credited mainly to the intense curiosity of the Arab scholars. They not only wanted to learn, but they also wanted to spread their knowledge, encourage experiments, and keep accurate records for posterity.

Arab alchemists improved laboratory equipment and identified many compounds. Words such as *alkali, elixir,* and *alcohol* were coined by the Arabs. Like the Chinese, perhaps even because of Chinese influence, many Arab alchemists linked medicinal benefit with transmutation. Some alchemists made drugs that were either reputed to or actually did effect cures.

The most famous alchemist of the Arab world was Jibir ibn-Hayyan or "The Great Geber" who lived in the eighth century A.D. Known throughout the civilized world, he was credited with ninety inventions and discoveries.

After his father, a poor druggist in a small town on the Euphrates, was executed for plotting against the ruling caliphs, young Jibir moved to Bagdad. There he devoted all his waking hours to the study of astrology, medicine,

The artist who did this portrait of the Great Geber obviously thought of him as a saint of contemporary science.

and mathematics. His knowledge and wisdom eventually won him the position of court scholar under Harun al-Rashid, ruler of Bagdad. Jibir, with full use of the palace library and laboratory, launched his fabulous career of discovery. Nitric acid was just one of many important compounds that Jibir discovered and described. He wrote volumes about his work and observations.

Even long after his death, the writings of Jibir continued. But this was no mystery. Many alchemists and scholars of lesser fame were simply using Jibir's name to draw attention to their own work.

Thomas Aquinas

While alchemy flourished throughout the Moslem world, it was slow to take hold in Europe. This was mainly due to the Christian church, which not only suspected anything coming from the "heathen" Moslem world, but also forbade dealing in any secrets or mysteries outside the faith.

A few Christian scholars ignored the ban, met with Moslem alchemists, and enjoyed the open cooperation and friendly exchange found among scientists of the world today. But despite Arab encouragement, alchemy remained little known in Europe until late in the tenth century A.D. The biggest boost for the young science came with the establishment in 1181 of the University of Montpellier in Southern France. Some deceptions were still needed to avoid church censure. Although all aspects of the alchemy were taught, the word was never mentioned. As far as the church knew Montpellier was a medical school, nothing more.

There, however, Roger Bacon, Albert Magnus, and Thomas Aquinas received their early training in science. Other famous Montpellier alumni included the astrologer, Nostradamus, the theologian, Erasmus, and the writer-physician, Rabelais.

The church did not relax its rules. On the contrary, as alchemy gained followers, the church increased its vigilance. In the thirteenth century, Roger Bacon, accused of possessing magic powers, lost his precious library to church confiscation and was later imprisoned for heresy. It is remarkable that he continued his work, advancing the

Albertus Magnus

study of both medicine and physics. His inventive mind ranged from such visions as powered flight to a practical way to process potassium nitrate into gunpowder.

As alchemy caught hold in Europe, the division between the theorists and the practitioners widened. To the theoretical alchemists or "dreamers" as they were disdainfully called, the discovery of a natural law that would explain some facet of the universe was far more important than a lump of gold.

The theorists, of course, were hampered by centuries of ignorance and wrong thinking. Although some Greek thinkers had suggested atomic structure of matter, there was another theory, also of Greek origin, that dominated all scientific thinking.

This was the four element theory originally expounded by Aristotle in the fourth century B.C. It stated that the only elements of matter were air, earth, fire, and water, and most things were made up of just two of these things with one holding dominance over the other. A liquid, for instance, was composed chiefly of water. A solid was composed chiefly of earth.

A profound thinker, Aristotle's great and lasting fame was justified, but several of his theories about earth and universe were wrong. Because of his fame, however, these theories persisted unchallenged for many centuries. And it came to be that any time-honored theory, no matter what its origin, should not be challenged. Thus the notion that gold grew underground remained unquestioned for centuries.

Gold, after all, was found in the ground, and the veins of gold running through rocks often looked exactly like the roots of a growing plant. It seemed that some regions were rich in gold because the wise people who once lived there had sown the ground with golden "seeds." Over the years, the seeds had grown into rich deposits.

Other theories, just as erroneous, influenced and misled the alchemists. It was thought for one thing that the various metals had affinity with various heavenly bodies. Tin was associated with the planet Jupiter, iron with Mars, copper with Venus, and mercury, of course, with Mercury. Silver was associated with and influenced by the moon while gold was closely related to the Sun.

It was motion of the heavenly bodies that caused changes or transmutations in metals. The sun could change a baser metal into gold and also change gold back into a base metal again.

Although the practical alchemists called them dreamers, the theorists had a strong influence on experiments going on in the laboratories. If gold grew underground, it was only logical that under the right conditions, it could be made to grow in the laboratory. The practitioners also timed their work according to the orbit of the planets in order to favor transmutations. Because they often used bellows to force more heat from the flames of their furnaces, the practitioners were referred to jokingly as "puffers." In any case, they made many great contributions to science.

Sir Isaac Newton (1642–1727), who developed the laws of gravity and invented an early form of calculus, knew all about the work of the puffers and even performed some of their experiments.

Although more famous for their theological advances, Saint Albertus Magnus (1193–1280) and Saint Thomas Aquinas (1225–1274) were both puffers before they took holy orders. Aquinas, always troubled by ethical questions, wrote in one of his many papers on alchemy that manu-

This drawing, based on a seventeenth century painting, shows a dedicated puffer at work in his laboratory.

factured gold was just as good as natural gold. He said he had used both and found one just as good as the other.

Alchemy could be dangerous. More than one puffer blew himself and his laboratory to oblivion. Alchemists worked with poisonous compounds, the worst of which were those containing mercury. Breathing mercury vapors could be, and often was, fatal.

Another danger that was more subtle than vapor came from the patrons. Strange to say, while the church banned alchemy, princes and kings encouraged the art. The rulers of the Dark Ages never had enough gold, and they would do anything to get it. They were usually eager to hire anyone who claimed to know the secret of growing gold. But it was risky to disappoint a king. More than one alchemist who failed to produce the promised gold was sent to the dungeon. Some were tortured. Others were beheaded, hanged, or strangled according to the fashion of the day.

Despite the dangers, there always seemed to be an ample supply of alchemists looking for patronage from kings or princes. Most of these alchemists were frauds. All through its history, it seems that alchemy has had more frauds than sincere scientific researchers.

An old Arab saying had it that alchemists were better at finding fools than at finding gold. And Arabs loved to tell the story of the foolish sultan who gave an alchemist a great deal of gold in the belief that the alchemist could make it increase overnight. Of course, next morning both gold and alchemist had vanished.

One group of frauds preyed on the poor as well as the rich. These were the "medicine men" who sold concoctions of gold as cures for every known ill. Unfortunately, when those concoctions actually did contain gold, it was often contaminated with mercury. The cure in these cases did more harm than good.

Amateur alchemists, though they never created gold, did touch off domestic disputes. The wife in this contemporary drawing is probably objecting to the strange smell coming from her kitchen.

Kings were not always gullible. After wars with England had drained his treasury, Charles VII of France had alchemists blend copper and silver to produce a false gold. The king promptly ordered the alloy molded into coins to pay his army. The French soldiers soon discovered the deception and lost trust in all government coinage. The French were not alone. English soldiers, already deceived by counterfeit coins, were insisting on payment in real gold.

Alchemists were blamed for a treasure horde discovered after the death in 1612 of Rudolf II, ruler of the Hapsburg Empire. When officials said that three tons of silver and four tons of gold, never recorded by the royal treasurer, had been produced by alchemists, all frauds encouraged the story. Even the church played this game. When Pope John XXII died in 1334, the heaps of gold found in his treasury were hard to explain until church officials thought of the alchemists.

Today, we are likely to laugh at the gullibility of our ancestors, but fraudulent alchemists have found believers even in our own century. Following World War I, Franz Tausend, a German plumber, made a fortune by selling stock in a gold-making scheme based on a "recently revealed" formula of the alchemists. Tausend worked his magic during the dawn of the Atomic Age.

Today, thanks to the Atomic Age, we can excuse such gullibility by saying that the claims of the alchemists were true. Gold can be made in the laboratory, and a certain kind of gold can be used to cure certain ills.

Laboratory gold was first produced in 1936 at the University of California in Berkeley. It took a combination of iridium and platinum, both rare metals, and an electrical jolt of thirty-eight million volts for the Berkeley scientists to produce gold. Of course, it was too expensive to be practical. The electricity needed to trigger the neutron bombardment that produced the gold would have kept the lights in a large city glowing for several months. And the gold itself was useless.

As soon as the power was turned off the artificial gold began losing electrons. As the hours passed, its atomic structure changed. The gold decayed into baser metals.

Using a different method, scientists at Columbia University in New York produced a more stable gold. They put platinum in a glass tube containing radium. Atomic particles given off by the radium attached themselves to the platinum and gradually changed it into gold.

Although the Columbia gold also decayed, the rate of decay was much slower than the rate for the University of California gold. This slow decay rate has proved extremely useful in the treatment of cancer, arthritis, and other diseases that respond to radiation treatment.

A small lump of laboratory gold implanted in diseased tissue will give off just the amount of radiation needed to halt the spread of the disease without harming surrounding, healthy tissue.

The ancient alchemists who were sincerely trying to benefit mankind would have approved of this development. In fact, the origins of most phases of modern science can be traced to the serious alchemists of the past. They did more than anyone else to overcome scientific ignorance and advance our knowledge of the world. Without them, we would still be living in the Dark Ages.

Here then is a case where the quest for gold has been a blessing to us all. It was not always the case.

Chapter Eight

The Curse of Gold

They drowned by the hundreds. Weighed down with gold they would not give up, the fleeing soldiers simply could not swim. They and their booty sank beneath the surface of Lake Texcoco never to be seen again.

Some five hundred Spanish soldiers were lost largely through their lust for gold. The event was the opening tragedy in Spain's conquest of the New World, and no event illustrated the future so well. Spain's lust for gold would prove a curse.

Actually, Spain was doubly cursed by gold.

Her native gold, as we have seen, brought enslavement by the gold-hungry Romans. When the resource gave out, Spain was left so poor and so drained of resources that it could not prevent invasion and long occupation by the Moors.

Even after the Moors were driven out, Spain remained poor. But then came the discovery and conquest of the Americas. Thanks to Christopher Columbus, Spain controlled the conquest from the beginning. Gold flowed into Spain.

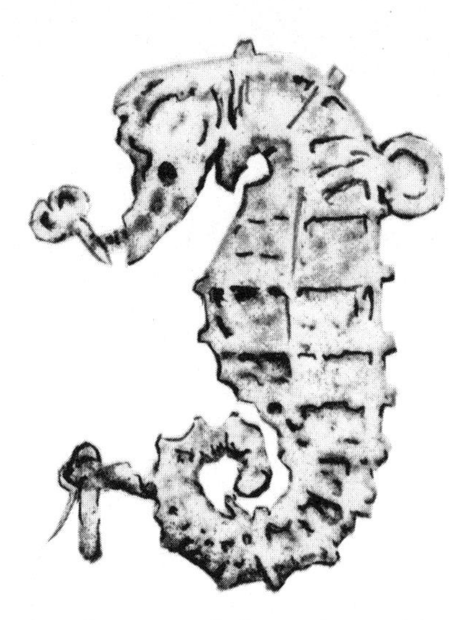

A seahorse, made by native smiths in what is now Panama, survived the conquistador's great melt-down of gold artifacts.

The first American gold, already mined, smelted, and worked into statues, jewelry, and a myriad of other exquisite ornaments by native craftsmen, was simply stolen by the Conquistadors. They melted the gold down, poured it into ingots, and shipped it to Spain.

Between 1492, the year of discovery, and 1550, when the native artifacts began to give out, from five hundred thousand to eight hundred thousand pounds of gold were shipped to Spain. And by 1810, after intensive prospecting and mining, the New World had yielded some two and one half million pounds of gold.

These estimates, however, do not account for the great sums of gold that were taken but never officially tallied on government records. It is enough to say, however, that a great treasure in both gold and silver fell into Spanish hands.

It produced no lasting benefit. Spain today is still one of the poorest nations in Western Europe. Why? What happened to all the gold?

There were many errors made in exploiting the New World, but behind them all was a common misconception about wealth. Like the Romans, the Spanish looked upon gold as wealth rather than a symbol of wealth. And like the Romans, Spain built its economy on gold.

We know today that wealth is best measured by the ability to produce. Rich farmland, forests, and natural deposits of minerals and fuel, combined with skilled labor are the true foundations of wealth. These basic resources produce both life's necessities and life's luxuries. While gold as a symbol does facilitate finance and trade, it produces nothing.

Perhaps if Spain had used its New World gold to build factories and train craftsmen, its story would have been different. But almost all the gold that flowed into Spain was spent to buy goods manufactured elsewhere, particularly in England and the Netherlands. Thus, the English and the Dutch benefited most from the Spanish conquest.

Spain lacked leadership, and again, gold was largely to blame. The New World, with its lure of gold, drained the country of its youth. No man with ambition, imagination, and energy wanted to stay in Spain when the Americas beckoned with so much promise.

Spanish royalty, instead of giving wise leadership, made bad laws and set poor examples. The Spanish throne, even though claiming a fifth of all gold discovered, never had enough to pay for royal luxuries.

The demand for the "king's fifth" was a classic example of bad law. It simply could not be enforced. Anyone willing to bribe a few key officials could smuggle gold into Spain and avoid paying the royal tax. Thus, the throne received much less than its fifth, perhaps just a tenth of the New World treasure.

The crown was unable to cope with inflation. Perhaps no government could have handled the problem created by too few goods and too much gold. Whenever the fleet arrived in Spain with ships weighed deep with treasure, the price of everything soared to higher levels.

Even when the fleet was late or lost to storm, causing a brief shortage of gold, credit transactions continued at the same high prices. Members of the aristocracy, following the example of their king, frequently went deep into debt on the anticipation of new gold shipments.

Eventually, merchants lost faith in all members of the nobility, including the royal household. When the king himself could not pay his debts what hope was there for the country?

The situation inevitably led to the rise of a middle class—the merchants, craftsmen, and manufacturers who provided services and goods. Most of the gold and silver stolen and mined in the New World during the age of conquest eventually fell into the hands of middle-class Europeans. This brought a profound social change, for as it gained wealth, the middle class gained power.

Control of human destiny was gradually transferred from the nobility to the merchant class.

Unfortunately, Spanish kings retained their power long enough to impede beneficial development of the New World. In fear of losing its royal share, the crown put so many controls on colonial mining that it was almost impossible to wield a shovel without breaking the law. As a consequence, laws were ignored and most mines produced great sums of contraband gold never recorded by the king's agents.

Spain's trade policies were particularly shortsighted. The crown ruled that its colonies could purchase no goods from foreign merchants. This gave the colonies the choice of buying goods at inflated prices from Spanish traders or breaking the law by dealing with English, French, or Dutch merchants. As you can guess, most colonists broke the law.

So the net effect of Spain's policies was to encourage trade with foreigners. The English merchants took particular advantage of the situation. And some of these merchants became part-time or full-time pirates who played no small role in challenging Spanish sovereignty in the New World.

Spanish policy saw to it that the colonies remained poor. Although much gold and silver passed through the

The alligator god of the New World may have put a special curse on the conquistadors. This little figure was molded with the lost wax technique.

towns and villages, little of it remained. Defenses of even the most important port cities were always inadequate. English pirates sacked and burned these port cities, sometimes cleaning out treasure houses where huge sums of gold were waiting shipment.

By the middle of the sixteenth century, Spanish treasure ships were forced by threat of piracy to sail in convoy under protection of armed galleons. By the year 1600, the convoy or treasure fleet numbered one hundred fifty vessels. With that many ships there were bound to be stragglers who were easily taken by enterprising pirates.

Actually, storms took far greater toll than pirates. Hurricanes often scattered the fleet, and many treasure ships went down. The treasure from ships that crashed on shallow reefs could usually be recovered by divers, but ships that sank in deep water were lost forever.

By far the worst of Spain's colonial policies was actually a lack of policy regarding native populations. Christians of the day, unable to explain how people could have wandered so far from the Garden of Eden, suspected that the Indians of the New World were not related to Adam and Eve. Therefore, logic had it that these strange people weren't human.

Even though Columbus had praised the Caribs, the first natives he met, as friendly and gentle people, the Spanish who followed had little sympathy for Indians.

Spain's first gold-seeking venture on the Island of Hispaniola, now shared by Haiti and the Dominican Republic, practically eliminated the Caribs that Columbus had praised.

Originally, some three hundred thousand Indians lived on the island. The first Spaniards enslaved all natives they encountered and ordered them to search the streams for nuggets and gold dust. When this approach failed to bring in gold fast enough, the Spanish adopted a system of tribute. Each native, on the penalty of death, was required to bring in four hawk's bills filled with gold each year. Just one hawk's bill was difficult to fill, particularly on Hispaniola where gold was actually scarce.

The Spaniards thought they could simply sit in their huts and watch the tribute flow. Instead they had to spend their days tracking down and killing fugitive natives. Natives who were not killed either died of starvation or succumbed to epidemics introduced by the Spanish. By 1519, just two thousand of Columbus's friendly Caribs remained on the island. By 1550, the number had dropped below five hundred.

The Conquistadors did little to improve on this record. Fabulous deposits of alluvial gold found in Panama, Venezuela, Ecuador, Mexico, Bolivia, Peru, and Chile were relatively easy to mine. Some river banks actually glistened with gold. A fortune could be gathered in a few days. Just the same, the European gold seekers persisted in enslaving the natives to do all the work.

In 1494, just two years after discovery of the Americas, Portugal and Spain ratified the Treaty of Tordesillas, thereby dividing all new lands between them. Brazil, India, and Africa went to Portugal. All the rest went to Spain. Although the treaty was endorsed by the Pope, it did not have approval from any other European country. English, Dutch, and French merchants and pirates were quick to challenge Spain's control.

By the time it became clear that the treaty could not be enforced, the foreign nations had established colonies along the Atlantic coast of North America. The English even maintained a trade and pirate center at Port Royal on the southern shore of Jamaica, right in the heart of the Caribbean. Although Port Royal was eventually destroyed by earthquake, the North American colonies, after some reversals, began to prosper.

In many ways, it was fortunate that there were no early discoveries of gold in the northern colonies. The settlements there soon became self-sufficient and sometimes almost ignored by their parent governments.

The Spanish colonies, hampered by royal rules, were not encouraged in self-sufficiency. Even when farming did begin, the major crops, like the gold, were sent to Spain in exchange for high-priced trade goods. The colonies did not break away from Spain until long after the gold had given out. It was gold, gold, always gold that led the conquest and kept Spain's interests alive.

Hernán Cortés, the first to lead an expedition into the highlands of Mexico, was hungry for gold. Although short of formal education, Cortés was clever, wise in the workings of Spanish politics, and apparently very persuasive and diplomatic with everyone he met, including the New World natives.

Although his 1519 expedition was authorized by the governor of Cuba, Cortés broke his ties with Cuba soon after reaching the Mexican coast. At the shores of Yucatan, where he made his first landing, the friendly Maya presented the Spanish captain with many gifts, including an Aztec princess.

Soon known as Doña Marina, the girl became Cortés's mistress and official interpreter. She was a lucky find. Without her, it is not likely that Cortés could have led his small band of six hundred men safely into the hostile interior. Cortés soon had more luck.

When he landed near what is now Vera Cruz, Doña

Marina learned that the band of cannibals holding the coast was ready to challenge Aztec rule. The girl persuaded one thousand tribesmen to join Cortés on the inland march.

From Doña Marina, Cortés learned more about the Aztecs and their legendary capital city, Tenochtitlan. It was built on islands in the middle of Lake Texcoco, a large body of water surrounded by lofty mountains.

From Tenochtitlan King Montezuma ruled all of Mexico, demanding and getting yearly tribute in gold, jewels, slaves, and other valuables from every tribal village in the land.

The Aztecs worshipped many gods. The sun god was thought to die each night and would not rise for a new dawn without first being revived with human blood. Sacrifice was thus a daily necessity. Slaves and captives also had to die to appease many other gods. We are still not sure how many Aztec gods there were or what they represented, but one of them had white skin and wore a beard.

By more good fortune for Cortés, the white god, according to Aztec legend, had left Mexico many years before, promising to return and bring vengeance against all who had forgotten to worship him.

Soon after hearing of Cortés's landing, Montezuma sent ambassadors to the coast to find out if the god whom everyone feared had returned. The ambassadors, though unsure of Cortés's divinity, took no chances. They bestowed gifts of feathers and gold upon the white stranger and begged him to go away.

Cortés did not try to mislead Montezuma's men. He told them that he needed gold and that a powerful king had sent him to get it. He gave the ambassadors a Spanish helmet and asked them to have their chief fill it with gold

This contemporary drawing shows Doña Marina standing nearby as Cortés, on a throne, receives the Aztecs' golden tribute.

and send it back to him. When the ambassadors left, Cortés led his men deeper into the jungle. He was already convinced that the stories about Montezuma's golden city were true.

The ambassadors returned with more lavish gifts, including the helmet brimming with gold, and strengthened the determination of Cortés and his men. The "White God" would not go away.

Even if Cortés were not divine, Montezuma had good reason to fear him. Aztec rule was harsh. Revolt needed little encouragement. Cortés would gather more and more support from local chieftains the farther inland he marched. And this indeed happened.

When, after a hard march of three months, Cortés and his men finally looked down into the Valley of Mexico, they stood in the company of many native warriors. Below them lay Tenochtitlan, a golden citadel of three hundred thousand people, a city that seemed to float on a glistening lake. The sight was as splendid as any view of a European capital.

As he led his small army down to the shores of the lake, Cortés, who had seen nothing grander than a native hut since arriving in Mexico, could hardly believe his eyes. Linked to shore by a long causeway, the island city seemed to grow larger and larger. When he reached the entrance to the causeway Cortés stood looking up at rooftops and glistening temples.

Montezuma, riding a golden litter borne by servants, came out to welcome the visitors. The emperor wore cotton garments dyed in many colors and topped with a brilliant mantle of tropical bird feathers. He also wore golden jewels and a golden crown. Even his sandals glistened with gold. He invited Cortés to bring his men into the city.

Every hospitality was provided. Servants bathed the trail-weary soldiers. Meals were served on golden platters. Fresh clothing and many gifts, including golden ornaments, were bestowed on them. But it did not take the Spaniards long to see that they were captives in the city. Drawbridges had been pulled up after them, and though they were free to roam the city as they wished, there was no escape.

As soon as he realized his predicament, Cortés took action. He ordered his men to seize Montezuma. The emperor would remain hostage in his own palace, Cortés declared, until all Aztec gold was turned over to the Spaniards.

The strategy worked. If anything, it worked too well. The amount of gold controlled by the Aztecs was far greater than Cortés thought existed in the entire world. It came to Tenochtitlan in seemingly endless flow. And the people who brought it did not seem to appreciate its phenomenal value. Some actually appeared to like the idea of giving gold to save their chieftain.

Montezuma himself provided guides to take some of Cortés's men to the mines where much of the gold was being unearthed. And Montezuma explained the system of tribute that provided gold from more distant reaches of empire.

Months passed and the flow of gold never slackened. Most of it had been worked into plates, bells, headdresses, lip plugs, pendants, collars, rings, and other gleaming ornaments. But there was some raw gold including twenty large jugs filled with gold dust.

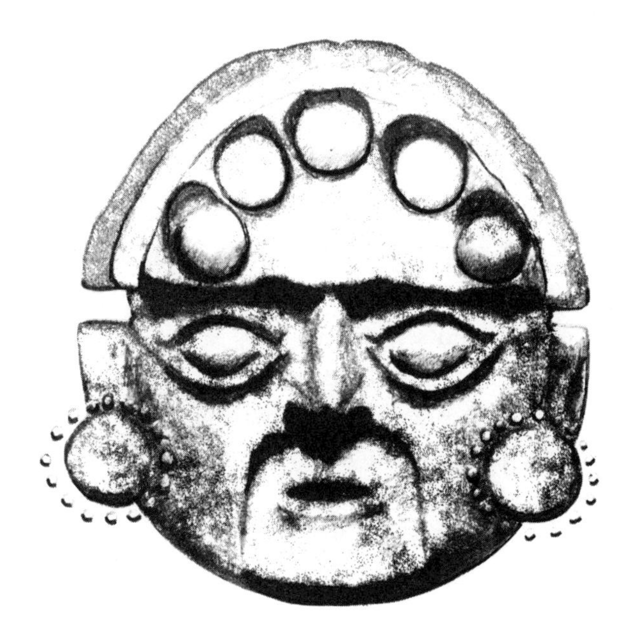

The Aztecs, along with many early people of the Old World, worshipped the Sun. This golden sun disk suggests that the Sun was a stern god in Mexico

A crew of Cortés's men worked daily at a furnace, melting down the skillfully worked gold and pouring it into bars. Each bar was stamped and numbered. Most of the spoil was divided among the soldiers, but Cortés, with Montezuma's help and approval, had some gold shipped directly to Charles V in Spain.

Winter passed into spring before Cortés had to face the first of his troubles.

The governor of Cuba, hearing of the direct shipment of gold to Charles V, realized that Cortés was acting as an independent agent. This violated the original agreement, which gave the governor control of all treasure from the expedition.

A fresh expedition, quickly organized in Cuba, was sent to Mexico with orders to strip Cortés of command. Warned of the expedition, Cortés left eighty men in charge at Tenochtitlan and led a force to the coast where he surprised and captured the governor's men.

With a few diplomatic words and the promise of riches, the governor's men switched their allegiance to Cortés and joined him in the return to the Aztec capital. There Cortés found serious trouble.

Aztecs, who had finally lost patience with the Spaniards, held the eighty-man garrison under siege. And when Cortés broke through to lift the siege he learned that Montezuma, in trying to stop the uprising, had been stoned to death by his own people.

Now the Spaniards realized they would be lucky to escape alive. They were outnumbered. No hope of negotiation remained in the hostile city. And there was just one exit—the causeway with its several drawbridges.

Just the same, they had to escape. Cortés selected the night of June 30, 1520, and ordered the construction of portable ramps. They would be used to bridge the gaps in the causeway. Division of the treasure was completed. Most soldiers bent their gold bars into circles so they could be worn like rings around arms or legs.

Rain began falling early on the fateful night. Cortés hoped the bad weather would further conceal his escape, but the Aztecs, well-informed of the Spanish plans, were waiting. The native soldiers pursued Cortés's men onto the causeway where they could be attacked from both sides by natives in canoes.

The Spaniards panicked. Most of the ramps were dropped. Blinded by night, terrified soldiers tumbled into the gaps and sank like stones. A few kept their heads enough to shed their gold and armor before jumping into the water. Some of these managed to swim to safety.

But most of Cortés's force was lost on what has become known as *La Noche Triste* or "The Night of Sorrow." Those not speared or drowned were captured for sacrifice. Almost all of the Aztec gold was lost in Lake Texcoco. One who managed to escape with his gold confessed that he walked across at least one gap on the backs of his dead or dying comrades.

Cortés, with a survival force of about a hundred men, retreated for six days. Then, on July 7, tired, hungry, and dispirited, they were overtaken by a horde of some two hundred thousand Aztecs.

Miraculously, the desperate Spaniards put the native forces to rout. The victory was all Cortés needed to begin undermining Aztec rule in Mexico. Tribal chieftains, long weary of the cruel demands of the Aztecs, gave their allegiance to Cortés. Chieftains who showed any reluctance were soon persuaded by force to join the revolt.

The Aztec influence dwindled so rapidly that Cortés

was soon able to march back to Tenochtitlan with scant opposition. He entered the city in the summer of 1521, little more than a year after the "Night of Sorrow." Then Mexico, with all its riches, belonged to Spain.

Like most New World Indians, the natives of Mexico valued gold for its beauty, workability, and any spiritual quality that might be given to it. Even though gold was hoarded and gold was demanded as tribute, it was not used in trade. Most raw gold was given to smiths to convert into ornaments and jewels.

Although only a few samples of native gold remain, we know that the master smiths in Mexico at Cortés's time were the Mixtecs. They revered gold as a metal made by the gods. The careful craftsmanship reflected this reverence. Their style stressed detail rather than bold expanse.

This alligator pendant made by Panamanian smiths is almost six inches long.

The Mixtec smiths could shape gold with beating or casting. They could join two or more pieces with soldered seams that the naked eye often could not detect. When the Mixtecs were absorbed by the Aztecs, captured smiths were forced to work gold and teach their skills in Tenochtitlan. Thus the Mixtec style dominated when Cortés and his men began melting ornaments into gold bars. If the work had been saved, it would now have value far exceeding the value of the gold itself.

The style of Peruvian gold, if reports from the conquest can be believed, was much more lavish than the Mixtec style. The Incas of Peru seem to have had access to a limitless supply of gold, and the goldsmiths there did not use the metal sparingly.

In Cuzco, the Incan capital, the inner walls of temples and palaces were sheeted with gold. Outside, full-sized golden animals stood among plants fashioned in gold from branch to leaf. Gardens of some Incan nobles had rows of golden corn. The Incan smiths could combine gold and silver into lifelike mice, butterflies, lizards, and snakes. Golden vines seemed to sprout silver leaves.

Francisco Pizarro, who conquered Peru, followed the example earlier set by Cortés. Pizarro captured the Incan chief and held him for ransom. Natives filled a twenty-seven-by-twenty-two-foot treasure room with gold to buy their chief's freedom, but Pizarro double-crossed the Incas. He killed the chief and marched into Cuzco to capture the golden city.

Like Cortés, Pizarro had all the golden statues and ornaments melted down for easy shipment back to Spain. The Incan Empire collapsed and Peru was left no better off than Spain is today.

Chapter Nine

Golden Brazil

The gold rush in Brazil, the first and most enduring in the New World, had its beginnings when Portugal and Spain divided the yet unclaimed regions of the world.

In the Treaty of Tordesillas, Portugal gained possession of Brazil because no one knew it was there. The treaty put all lands lying one hundred leagues west of the Cape Verde Islands in Spain's possession and all lands east in Portugal's.

Spain thought the division would give it all of South America, but the explorers of the day had not yet discovered that a large hump of the continent lay east of the boundary line. Some historians now suspect that the Portuguese who signed the treaty may have heard rumors about the Brazilian hump, but it is most likely that both parties in the treaty acted with an equal measure of ignorance.

It was not until 1501, after the treaty had been in effect for seven years, that the true shape of South America became known to the world. In that year, Amerigo Vespucci, the Italian navigator commissioned by the Portuguese to find a passage to the East Indies, sailed west until he found land.

He explored the coast thoroughly, claimed the land for his employer, the king of Portugal, and became so closely associated with the New World that "America" was derived from his first name.

Vespucci did not bring home many favorable reports about Brazil. Although the natives were friendly, he thought little of their judgment. When offered gifts, the Brazilians usually chose mirrors or combs rather than anything of value. They even seemed to prize bird feathers above gold.

The natives told of huge deposits of gold in the jungle streams, but Vespucci did not investigate. He thought the stories were lies.

Later explorers did nothing to change Vespucci's opinion. Though vast in area, Brazil offered few incentives for settlement. Thus it was that the first to settle in any number were those driven to the new land by persecution at home.

They were originally Jews. The inquisition had forced them to convert to Christianity, and when persecution continued, they fled to Brazil. They established colonies along the coast and tried to make a living by exporting salt dug from natural deposits and the fibers of a tropical tree, which made a red dye. Although Brazilwood eventually became a valuable export, the first settlements were poor and largely ignored by the homeland.

When plantations began producing sugar cane and tobacco for export, colonial economy improved, but the inland regions of the vast territory remained unexplored and undeveloped. The first to move inland from the coastal settlements were not Europeans. They were half-breeds.

They called themselves Paulistas because most of them grew up in the countryside around São Paulo.

It had been common for the men of the early settlements to take native wives and produce a great many offspring. So half-breeds were plentiful, but they were not accepted socially by either the natives or the Europeans. Many of these outcasts followed the inland trails, took on native ways, and turned their backs on colonial society entirely. The Paulistas searched for game, rare metals, and precious stones. They also raided jungle villages in order to take captives and sell them in the slave market.

At first, slaving brought the most profit. Although the native slaves were not as hardy as those imported from Africa, a native could be bought cheaply without the heavy tax imposed by Portugal on imported workers. The Paulistas soon gleaned the best workers from the coastal tribes and were forced to range farther and farther inland on their raids, eventually reaching the base of the Andes Mountains. They were Brazil's first explorers.

Although their main interest was slaves, they often brought home small quantities of gold that had either been found or stolen from natives. The Paulistas sold their gold easily and at a good price, but for a long time none of them took up prospecting as a full-time enterprise.

All this changed in the middle of the seventeenth century when Portugal's gold-based economy faced ruin. The government was running out of the gold and silver needed to support trade with its colonies and other countries. Actually, a large part of the problem was due to shortsighted colonial policies.

Like Spain, Portugal forbade its colonies to trade with foreigners. But the high prices on Portuguese goods and high tariffs imposed by the government made trade with the homeland very expensive. Foreign trade, even though illegal, thrived. This foreign trade siphoned away most of the gold and silver that might have gone to Portugal if it had adopted a more liberal policy.

Instead of changing its policy, however, the government sent out a call for precious metals. In Brazil, the colonial government warned that the only thing to prevent economic failure would be the discovery of a new source of gold.

The situation was very much like that faced by our own energy-based economy. Without new discoveries of oil, today's economy faces collapse.

The call for gold sounded to the Paulistas like nothing more than a challenge. If the government needed gold, the Paulistas would get it.

There were no immediate big strikes, but by spending more time looking for gold and less time rounding up slaves, the Paulistas began putting more and more gold into circulation. In the early years, they had a monopoly. No Europeans knew the inland regions nor had the jungle survival skills that were part of the Paulistas' heritage. The situation changed suddenly, however, in 1682, when an outlaw discovered a "river of gold."

Manuel de Borba Gato, a Paulista wanted as an assassination suspect, had been hiding with friends near the Rio das Velhas, which flowed through some of the thickest jungle in Brazil. There they discovered river sands that glowed with golden dust. As they feverishly gathered the dust, Borba Gato and his friends agreed to keep the find a secret. But the news got out soon after the first of the das Velhas gold reached Rio de Janeiro.

Farmers left their fields, merchants locked up their shops, sailors abandoned their ships, and craftsmen dropped their tools. All headed for the gold fields. All were poorly informed and poorly equipped. It is remarkable that so many reached the gold fields alive.

Meanwhile, the Paulistas found gold in neighboring rivers. Rio das Mortes and Rio Doce, together with Rio das Velhas, became known as Minas Gerais or "General Mines."

The early gold seekers swirled the river sands in cone-shaped pans until the gold had settled at the bottom of the cone. Later, sluice boxes fed by crews of shovel-wielding miners were used to extract the river gold. Eventually miners began digging into banks up and down the river in search of buried veins of gold.

Loose earth brought up from the digs could be run through the sluices just like the river sand, but hard rock mining presented problems. It was heavy work just breaking the rock free. Then it had to be crushed and run through smelters. The equipment was costly. It required big investments of money. Some miners banded together, forming companies, and pooled their funds to buy the heavy equipment. Thus, most late arrivals at the gold fields, instead of striking it rich on their own, ended up working for daily wages with one of the big companies. It was dangerous work.

When veins of gold ran deep, so did the tunnels. They were vertical tunnels that dropped far below the river valleys and slanted or horizontal tunnels that led under the nearby mountains. Where veins led, the tunnels followed.

Unfortunately, the rock of the region crumbled easily. Cave-ins occurred often. Many miners were crushed to death. Many others were trapped without hope of rescue.

Life at the mines was never easy. The men lived in shacks. There were no rules of sanitation. Illness was common. While the mines drew men from the business community, the medical profession, and the clergy, they also drew a good share of murderers and thieves. Strangers could not be trusted.

It took twenty days to reach the mines, and all food and supplies, hauled over the long, rough roads, were terribly expensive. No miners had the foresight to grow their own crops.

If there had been laws, it's not likely that anyone would have enforced them. A dispute over a mining claim was often settled with the gun or the long knife.

To make things even worse, the Paulistas, who had long dominated the inland regions of Brazil, resented the newcomers. There were some fierce battles, and in some regions the Paulistas briefly held back the invasion of prospectors. But the call of gold was too strong, and the hordes of newcomers too large. Eventually, the Paulistas lost control of the rich fields.

Government leaders, who had hoped a gold rush would solve Brazil's economic problems, were badly disappointed. The government needed stable production of export goods from its plantations and forests, but the farmers and lumbermen had all gone to the mines.

Even in the coastal settlements, prices of food and other necessities soared. Those lucky enough to have gold could pay the high prices, but the majority of the people suffered.

In Portugal, Brazilian gold was spent quickly to buy imported goods at inflation prices. Again, the industrial nations of northern Europe reaped the profits.

At Minas Gerais only the shrewd merchants grew rich. Miners, never famous for frugality, spent their gold quickly on drink, women, and gambling. Actually, most miners were in debt to the merchants who provided tools, groceries, and other necessities at inflated prices.

If he also dealt in slaves, a merchant could grow rich quickly. The discovery of gold greatly stimulated importation of African slaves to Brazil. At first, the slaves were needed on the plantations to replace workers who had joined the gold rush. Later, many slaves were sent directly to the mines.

Smuggling, which the government had been unable to control before the gold strike, increased tenfold. Because of the tariff, foreign goods were always cheaper than Brazilian imports. And foreign traders were happy to accept payment in gold.

The foreign traders performed a valuable service for the miners because they smuggled gold out of the country. If a miner turned his gold over to a government-run smelting plant, he had to pay out one-fifth of it in royal taxes. Obviously, it was far better for the miner to turn his gold over to a foreign trader who gave full value.

Smuggling of goods and gold became big business. Some smugglers even maintained small armies to discourage interference from thieves or colonial agents. Rather than deal with foreigners, a few miners ran their own smuggling rings and used every trick imaginable to fool the agents. Dressed in priests' robes, miners could pass easily through inspection stations without anyone suspecting that the religious objects they carried were made of solid gold.

The law said all gold must be smelted. No one enforced the law. At the mines, nuggets or gold dust were used in every transaction. It was unusual to even see a coin.

It seems that the more the government tried to collect the royal tax, the greater the effort failed. In 1701, when

The search of gold took pioneers deeper and deeper into the Brazilian jungle and did much to open the country for settlement.

production at Minas Gerais was approaching full strength, the records show that just thirty-six miners paid the tax. In 1702, the tax was paid by just one man.

Remarkably, gold from the mines did reach Portugal, great shipments of it. Although almost all of it was smuggled gold, the shipments were recorded. They show a remarkable increase in production.

In 1699, gold imports totaled 1,598 pounds. In 1701, imports rose to 3,935 pounds, and in 1712, the figure jumped to 31,967. These, of course, are not full production figures for the mines. Because of smuggling by foreign agents, no one can guess how much gold was taken from Minas Gerais over the region's peak years of production.

We do know that English agents, or factors as they were called, cornered the most smuggled gold because the English had the largest variety and volume of trade goods to offer. The factors operated openly in all of Brazil's major towns and cities as well as at the mines.

For a time, colonial agents considered banning the English factors, but the country depended so heavily on English trade goods that the move would have been crippling.

Meanwhile, ceaseless work at the mines was having tragic consequences. Tunneling had given way to strip mining. And when strip mining proved too costly and slow, hydraulic mining began. A region that had once been lush jungle was being turned into a waste of raw sand and gravel. No one seemed to care.

The main worry of the miners was that the gold was giving out. But the Paulistas came to the rescue. These rugged pioneers kept making new finds.

By 1718, when Minas Gerais was all but paid out, new fields were being developed on Rio Cuiaba in the Mato Grosso district near the Bolivian border. In 1725, interest had shifted to the Goias district, and in 1734, miners were rushing back to new finds in Mato Grosso.

At each new field, the pattern was repeated. The government agents were unable to collect the tax. The merchants and the smugglers grew rich. The miners gained little from their hard work and high risks.

Rio Cuiaba was so remote that it took seven months of travel by jungle trail and river to reach the gold fields. Many died of disease or starvation or were killed by native arrows before they ever touched a grain of gold. If the government sent tax agents to Rio Cuiaba, we have no records of their success.

Even though new finds continued, Portugal benefitted little from Brazilian gold.

The Brazilian gold rush, however, never has really ended. And today, Brazil, as an independent nation, benefits substantially each time a new gold region is developed.

The most recent chapter in the Brazilian gold rush opened with the 1980 development of the Serra Pelada region in the heart of the Amazon River Valley. It was there, in 1982, that Clovis Tavares, a baker from São Paulo, found Brazil's biggest gold nugget. It weighed 615.6 ounces, and the Federal Savings Bank of Brazil, which now purchases all gold from prospectors, paid Tavares $394,273 for the giant.

Who can say how many more golden riches remain hidden in the jungles of Brazil?

Chapter Ten

Eureka!

Three men using nothing but spoons dug thirty-six thousand dollars in gold from cracks in a rock. A rabbit hunter poked a stick in the ground, hit rock quartz, and dug up ninety-seven hundred dollars worth of gold in three days. A man with a single scoop of his pan recovered twenty-nine hundred dollars in gold in less than ten minutes. In one spot a miner could spend the day shoveling and washing sand and earn five thousand dollars.

The stories were true. California sparkled with gold. In ten years, more gold was mined in California than was taken, by official record, during the entire Spanish conquest of the New World.

The story of the most famous gold rush in history also began with a treaty, not between Spain and Portugal but between Mexico and the United States.

It was no accident that the "discovery" of gold coincided with the signing of the Treaty of Guadalupe Hidalgo, which transferred Texas and California from

Mexico to the United States. Early American settlers in California had long known about gold, but they kept quiet about it.

It was well they did. If Mexico had learned of the rich deposits in the Sierra foothills, it would never have given up the territory so easily.

Actually, California Indians had discovered gold long before any Europeans arrived. The prehistoric natives put no special value on gold, but soon after the first explorers touched California's shores, Indians began offering gold dust for barter. True, no one was sure exactly where the gold came from or how rich the deposits might be, but many early stories and legends told of California gold.

The first published report appeared in a book written by an English mineralogist in 1816, thirty-two years before the "discovery." The book stated that California mountains were rich in gold. The Englishman's report was either not believed or not widely circulated. It failed, in any case, to trigger a gold rush.

The seven hundred or so Americans who had settled in California as guests of the Mexican government were mostly farmers and ranchers. Mexico welcomed them only because it needed production and stability in what had long been a remote and largely untamed territory. This guest-host relationship was delicate, and the settlers knew that Mexico had the right, at any time, to force them out. So if investments in buildings and livestock were to be preserved, nothing must upset the delicate relationship.

In this cautious atmosphere it is not hard to understand why a trapper who came out of the Sierra in 1826, with his pockets full of gold, never won fame. Persuasive settlers convinced the trapper that he must keep quiet about his find. Of course, news traveled by word of mouth

Sutter's Fort was the major trading post for settlers throughout California's Sacramento Valley.

among the American settlers. Everyone knew about the trapper's gold, but no one made a lot of noise about it.

A Swiss settler, Johann Sutter, had strong reason to keep down the talk of gold. He had invested all he owned in large holdings that included a big section of the Sacramento Valley and a ranch in the Sierra foothills. He hoped to use the lumber from the ranch at Coloma to build more homes and shops around the valley trading post that had already become known as Sutter's Fort.

It seemed, however, that every time he sent men to work at the Coloma ranch, they came back talking more about gold than lumber. In 1843, and again in 1844, large quantities of gold were collected at the ranch. It was all Sutter could do to keep the discoveries secret. By this time, war with Mexico was almost certain, and settlers in California were in a very awkward position.

The Bear Flag Rebellion and war with Mexico in 1846 did little to clarify the position. It was not until the end of the war drew near that the settlers began to feel confident. By the close of 1847, it was obvious that Mexico had lost its hold in California.

Sutter, however, remained cautious. Early in 1848,

James W. Marshall, hired to build the long-planned sawmill at Coloma, came down from the hills to tell Sutter that there was gold on the ranch. Impatiently, Sutter replied that he knew all about the gold. Marshall could do everyone a favor by keeping quiet about it. At the same time Sutter urged Marshall to return to the ranch and finish the mill.

It never happened. At the ranch, Marshall and his crew panned gold from dawn to dusk, knee deep in the American River. Workers from neighboring ranches soon joined them, and before long, the ranch was teeming with prospectors. Sutter began selling shovels and picks to gold seekers who had come all the way from Monterey and San Francisco (then called Yerba Buena). The cautious Swiss settler soon learned to his amazement that these men were headed for his ranch on the American River.

And the American, it turned out, was just one of several gold-rich rivers. In rapid order, finds were made on the Yuba, the Feather, and the Stanislaus. Mining camps came into being overnight, and they had colorful, spur-of-the-moment names—Timbuctoo, Ophir, Dutch Flat, Spanish Flat, Placer, Chinese Camp, Hang Town, and Rough and Ready. The first miners came from coastal towns where California's small American colony had been concentrated.

The treaty of Guadalupe Hidalgo was signed on February 2, 1848, but by the time news of it reached California shores, it seemed that the entire population had moved to the Sierra foothills. And by then prospectors were also coming from Hawaii, Mexico, South America, and even Australia.

Strangely, the early reports of California gold caused little stir in the eastern United States. Perhaps the sophisticated easterners had been wearied by tales of the fabled West. Or maybe it was too soon after the war with Mexico to consider new ventures. In any case, it was not until December 5, 1848, when President James Polk told Congress about the discovery, that the East woke up. By then, of course, California was under firm United States ownership.

So it was that the big rush to California did not begin until 1849, almost a year after the discovery. The Forty-niners had three choices. They could make the long, tough journey by wagon trail across the continent. They could take the expensive and risky voyage around Cape Horn, or they could sail to Panama, cross the neck of land there, and then hope to catch a boat to California. No matter what route was chosen, the trip took many months, and when the prospectors finally reached California, they found hopeless confusion.

The rush to the gold fields was so general that no one of ability remained in San Francisco to run the government or provide public services. The town, bursting with new arrivals, had no adequate police force or fire department. New houses went up so fast that it was hard to find your way through town from one day to the next.

By the end of 1849, what had been a small village at the shore of a natural harbor, had grown to a city of twenty-five thousand. And in 1849 alone, some forty thousand miners passed through the port city on their way to the diggings.

The harbor itself was forested with the masts of ships, rotting ships. All the crews had gone off to find gold, leaving no one to sail the ships away. Several holds remained full of cargo. There was no one to unload them.

Sailcloth was taken to build tents and canvas-walled

shacks. The smart merchants bought full ships, beached them and converted them into shops where customers bought the cargo at high retail prices. Other boats were beached for homes.

All during the 1850s, an average of thirty homes were added each day. The figure is misleading, however, because fires repeatedly destroyed the city. In one period of eighteen months there were six devastating fires. Of course, the city was quickly rebuilt after each blaze.

Though the shabby town did not suggest permanence, it was fabulously rich. In the ten years following Marshall's "discovery," five hundred fifty-five million dollars in gold passed through San Francisco's banks.

Prices were a scandal. Inflation forced the cost of a choice building lot from twelve dollars to ten thousand dollars. A shovel sold for ten dollars, and eggs for one dollar.

Prices were even worse at the mining camps. There shovels were selling for one hundred dollars each. An egg cost three dollars, and a sack of flour eight hundred dollars. But strangely there was more order and sense of purpose in the camps than in the city.

Everyone was busy getting rich or trying to get rich. There was little time for public service, politics, or law enforcement. The necessities took care of themselves. The rest was of little importance.

This is not to say that the camps lacked excitement. On the contrary, the small communities buzzed with each fresh bit of news, including those astounding stories of discovery.

Near Angel's Camp, a man known only as Raspberry was hunting deer when his rifle accidentally struck and broke a rock. The rock glistened with thick veins of gold.

In three days of solo digging, Raspberry reportedly dug up seven thousand dollars in gold.

In Grass Valley, George Knight stubbed his toe on a rock that turned out to be gold-bearing quartz, the only surface sign of a vein that would eventually make Knight one of California's richest men.

Near Carson Hill a miner found a nugget that weighed one hundred ninety-five pounds. He sold it for seventy-four thousand dollars.

Then there was the repeated tale of the miner's funeral that was abruptly terminated when gold was spotted in the open grave. The preacher joined the dead miner's friends in the frenzied dig. The miner was eventually put to rest somewhere else.

By some accounts, such tales could travel the two hundred-mile length of the Mother Lode—from Downieville on the north to Mariposa on the south—in just two hours. And as you might guess, such stories caused overnight migrations. A camp that had been bursting its boundaries one day might be empty the next after a story of a new find passed through.

When a new camp was established, the miners would usually meet to draw up rules for the diggings. The main business was to decide the size of a claim it would be practical for one man to work and how individual claims would be protected against trespass. There was little else to discuss.

There had never been a time or place so free. There was no taxation. There were no police, no government inspectors, no public officials of any kind. And there were no written laws.

There was, however, an unwritten law governing the camps. It rose from the character of the miners. Almost all of them were young, still in their twenties. Most had received good educations. They were full of enthusiasm and good will. They had faith in their fellow man. They believed in fair play.

Only a few dared break the miner's code of ethics, and they were dealt with harshly. A claim jumper or horse thief might be hanged or at least branded and banished without formality of trial. Murderers got the noose or an order to leave camp at once.

Sunday was a day of rest in the camps, and Sunday afternoon provided time for meetings if any were needed. Here disputes over claims were settled by majority rule, and criminal cases not already dealt with were heard.

Sundays also provided necessary social contact. Mining was a lonely life, and women were so rare that some miners claimed to have forgotten what the opposite sex looked like. The unusual miner who came to the diggings with a wife was blessed in many ways. One miner described "Mrs. R" as a "magnificent woman" who "earned her old man nine hundred dollars in nine weeks taking in washing."

Actresses who came to the camps with touring companies were given heroic welcomes and treated like royalty. One miner, who paid fifty dollars for an actress's slipper, remained the envy of camp long after the acting company departed. He drew a large crowd to his tent each time he displayed the slipper and reportedly turned down many tempting offers for the prize.

An 1850 census reported that the mining camp population was just eight percent female. The percentage of "ladies" was much smaller. At one time, a temperance lecturer tried to persuade the miners to import five thousand virtuous women from England, but the noble plan did not gather much support. Most unmarried women who came to the mines were prostitutes, busy prostitutes. One young woman reportedly earned fifty thousand dollars in less than a year and retired from the profession. Some did not fare so well. There was at least one case tried in the early courts of a prostitute who accused a client of paying her in watered-down gold dust.

In the early days there was remarkably little crime, but as the wealth of the California deposits became known, thieves, gamblers, and other social leeches moved into the camps. The innocent era faded into memory. Miners turned suspicious. The camps turned into wild towns. Crime turned into a serious problem.

Although there were some notorious robbers and murderers of the era, the more imaginative criminals dealt in gold mines. By loading a shotgun with gold dust and firing it at a rock, a "rich" lode could be created. Salting, as it was called, was a common practice.

Mark Twain, who wrote so colorfully of the West, described a mine as "nothing more than a hole in the ground with a liar on top."

The crooks who sold claims or stock in salted mines were hard to catch. Everyone knew, after all, that mines did not always produce as expected. That was the nature of the business.

The nature of the mining business in the Sierra foothills went through profound changes. In little more than ten years, the free-spirited, independent prospector gave way to the laborer who worked for wages under the direction of a company foreman. The changes in methods and conditions can best be symbolized by the changes in mining equipment.

Although the gold pan remained a handy tool for the prospector in testing river sands for "color," it was too

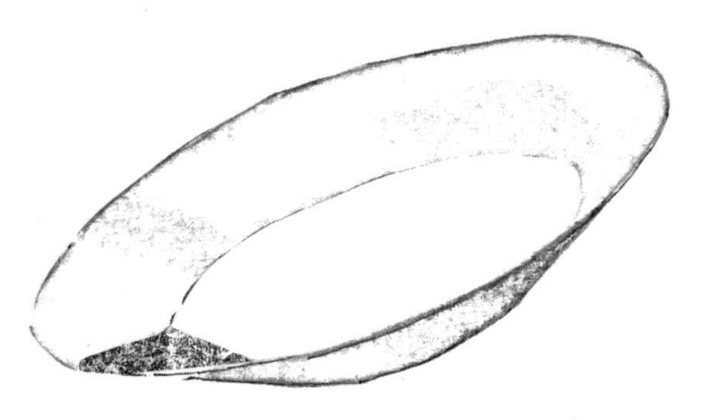

The typical gold pan was about eighteen inches in diameter.

small to extract much gold dust from a sandbar that might be forty to fifty yards long.

Independent miners soon adopted the cradle, which looked indeed much like a baby's cradle. The base of the cradle box, however, was slanted and lined with many ridges. The miner dumped a mixture of sand or gravel and water into the box and then rocked the cradle. The motion forced heavy material such as gold to settle to the bottom and catch in the ridges while the lighter material washed away.

The cradle, standard gear in the early days, was light enough to carry from one sandbar to the next, and if sands were rich, it could make just one day of rocking extremely rewarding. But as the rich sands grew scarce, the miners

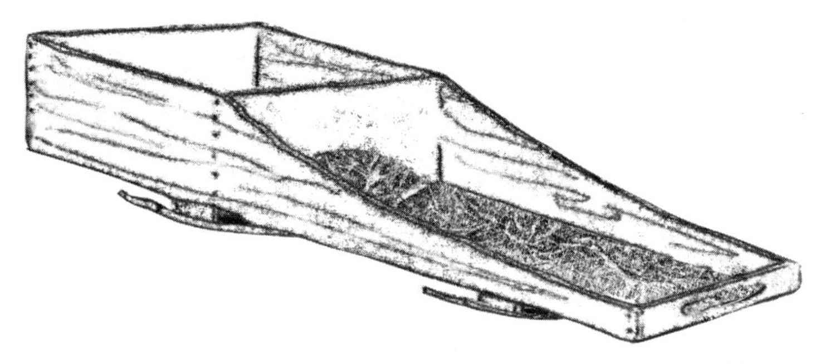

With a cradle and lots of luck an independent miner could process enough sand and gravel in one day to make a fortune in gold.

were forced to build sluices. This meant giving up their independence because the sluices were expensive to build and could not be operated alone.

Sluice design followed the conditions and contour of the river. The basic sluice, however, was a wooden trough that channeled water from a diversion dam, higher up the river, to the sands or gravels below. The gold-bearing material was shoveled into the trough and allowed to wash over a series of ribs further down the sluice. At the end of a working shift, the water was blocked off, and the miners carefully gathered up the dust that had collected behind the ribs. Earnings were shared equally among all the workers.

Sluice companies, the first cooperative ventures at the mines, were formed, disbanded, and reformed as needs and ever-changing expectations changed. Some companies made their members fabulously rich in just a few weeks, while others barely produced enough to pay a miner's expenses.

Not all miners joined companies. Many independents chose to go on rocking their cradles. Others who wanted to work alone took up pocket mining.

The pocket miner worked the river sands with his pan to trace "color" to its source. Sometimes his search led him far up a canyon, perhaps even into a dry gully. If successful, he located the vein of gold or even clusters of nuggets that had been, by erosion, "coloring" the river sands below.

Sometimes pocket miners came upon deposits that were so rich and so extensive that they could not be exploited without full-scale mining and smelting plants. Usually, to preserve independence, the lucky miner would sell his claim for the best price he could get and move on, leaving investors to form a company that would dig and process the ore.

While the cry of "Eureka!", Greek for "I have found it," may actually have been heard more than once in the California gold fields, there were thousands of miners who found more disappointment and sorrow than gold.

Gold ruined Johann Sutter. At the height of the rush, he could find no men to harvest his crops or milk his cows. Squatters not only overran his property but also clouded his ownership of some 71,221 acres of choice land. In 1855, when a court ruled that Sutter should be paid twenty-five million dollars to compensate for the loss of the land, an angry crowd of miners burned Sutter's buildings and tried to lynch the judge. Sutter was never paid a penny. Sutter's oldest son shot himself. A second son was murdered. The third son drowned. Sutter, who became a hopeless drunkard, died in poverty.

James Marshall, after failing to hold his early luck at Sutter's ranch, also took to drink.

Quartz or hard rock mining was hard, dangerous work. The usual method was to drop a vertical shaft and then dig several horizontal tunnels branching out from the bottom of the shaft. Some of these mines grew into huge excavations. The famous Empire Star Mine at Grass Valley had more than two hundred miles of tunnels.

Processing methods were so inefficient during the early years that ore had to be very rich to return a profit. Just the same, such men as Leland Stanford and George Hearst made fortunes through lucky investments in gold mines.

It was not until 1887, when the cyanide process was introduced from Scotland, that smelting the poorer California ores became profitable. Cyanide compounds, mixed with the crushed ore, picked up the gold, which was later precipitated with a bath in zinc compounds.

Meanwhile, with the cyanide process yet to be invented, miners concentrated on the alluvial gold. Sluice mining was limited by its dependence on a crew of shovelers to feed the sand and gravel into the trough. Sand and gravel could be processed only as fast as the crew could work. And that was not fast enough.

So the placer mines turned to hydraulic methods. Of course, they did not invent the method, but the California miners refined hydraulicking and applied it on a grand scale. Elaborate gravity systems of dams, aqueducts, reservoirs, and leather hoses delivered the water to the diggings under extremely high pressure. The water was "fired" through an iron nozzle that looked much like a long-barreled cannon.

And indeed, it was as destructive. It blasted away everything in its path. Forests were undermined. Mountains were leveled. And the tailings, after running through the sluices, were dumped into rivers. At the height of hydraulicking, the Sacramento River was so thick with mud that farmers could not use the water for irrigation. It clogged their diversion ditches.

This water cannon, known as the "Monitor," now stands on display in a small park in California's Nevada City.

As if hydraulicking was not damaging enough, the miners adopted still another way to overturn the earth—dredging. Huge floating dredges appeared wherever gold-bearing terrain was relatively level. The process could begin on a small pond or side stream, anyplace where the dredge would float. From this start, the dredge, sucking topsoil into its innards, moved relentlessly across the land. The gold was extracted by a sophisticated system of sluices inside. The silt was washed away while rocks and gravel piled up in the wake of the dredge.

Gold dredging, of course, added to the silt problem, and it ruined hundreds of acres of good farmland. The desrtuction left by the dredges one hundred years ago can still be seen—vast areas in the Sierra foothills lined with row upon row of sterile rock. They are a sad testimony to man's greed.

Finally, in 1884, the state legislature, responding to public outcry, put an end to hydraulic mining in California. But for some areas, the prohibition came too late.

Hydraulic mining and dredging were both expensive methods financed by big business. And when the focus returned to quartz mining with the newly introduced cyanide process, gold production remained in the hands of big business.

The independent miners, what few were left, became prospectors, hoping for rich finds they could sell to one of the big companies. But as the foothills grew crowded, it became hard to find areas that had not been searched.

The lone prospector took his pick and shovel and his gold pan to other regions of the unexplored West. He left behind him an era of spirited independence, enthusiasm, and good will that the world is not likely to see again.

Chapter Eleven

Where You Find It

The California gold rush was neither the first nor the last in the United States. Gold was discovered in North Carolina as early as 1799 when a boy hunting for a lost arrow found a large lump of gold.

The record says that the boy took the nugget home to his father, a man called Reed, who used it for a doorstop for several years before selling it. News of the sale brought prospectors to the Rocky River region where the nugget had been found.

For a time no major discoveries were made. Ten years of mining, however, produced one hundred fifteen pounds of gold, and then large nuggets began showing up again. One was found weighing twenty-eight pounds. And then, in 1828, a fifty-pound nugget was unearthed.

A minor gold rush began. Up until 1828, all gold minted in the United States came from North Carolina. In subsequent years other southern states with rivers draining the Appalachian Mountains contributed small amounts of alluvial gold. But by the start of the Civil War, after yielding some twenty million dollars in gold, the alluvial deposits were played out.

Meanwhile, in California, disappointed gold-seekers began leaving the diggings in search of richer regions even as new hopefuls were arriving. The wanderlust, so typical of this young breed of adventurer, drew miners to nearly every region of the west and sometimes far beyond.

"Gold is where you find it" was the byword. In most cases the byword simply encouraged false hopes, but occasionally it seemed to have magic power.

Edward Hammond Hargraves, who left his Australian ranch to seek gold in California had the magic. He found nothing but loneliness in California, and to make things worse, he was homesick. The Sierra foothills reminded him a great deal of the hills around his ranch back in New South Wales.

One day, while writing home, he was struck by simple logic. If there was gold in this land that looked so much like home, then there must be gold at home. He returned to Australia as quickly as he could.

On February 12, 1851, Hargraves, prospecting near home with a man called Lister, stopped in Summer Hill Creek, a tributary of the Macquarie River and tried the sand with his pan. He found gold dust, great quantities of it. Again and again, each panful yielded the gleaming stuff.

With great glee, Hargraves told Lister:

"I shall be a baronet, you will be knighted; my horse will be stuffed, put in a glass case, and sent to the British museum."

Hargraves, perhaps motivated by these dreams of honor, took the news of his discovery directly to the Colonial Secretary and thereby put Australia's gold rush under government control. The Australian government saw the discovery of gold as a cure for immigration problems.

While the country needed settlers, all it had been getting were convicts exported from the British Isles. Generally, the convicts were more of a liability than an asset, and making matters worse was the recent gold rush in California. Many of Australia's more reliable citizens had already left for California.

An Australian gold rush might turn things around. The colonial government appointed Hargraves commissioner of lands at twenty-four thousand dollars a year and made sure his discovery was well advertised. Soon an army of prospectors, each with a license to dig on government land, marched on the Macquarie.

The early finds were spectacular. A Bathurst blacksmith took eleven pounds in gold from a single hole. In just a few days of digging, a merchant from Sydney returned home with gold worth twenty-four hundred dollars.

Every able-bodied citizen in both Bathurst and Sydney headed for the "golden river." They called themselves "fossickers" a colonial term for ditch diggers. The *Bathurst Free Press* reported:

> *"A complete metal madness appears to have seized almost every member of the community. There has been a universal rush to the diggings."*

Reports of gold went beyond the seas. Communication and travel were slow, but just the same, by the end of 1852, the flood of immigration that the colonial government had long desired was well underway. The mines had yielded forty-five thousand pounds of gold, but better yet, the population of Australia had grown by three hundred and seventy thousand.

Clipper ships, which could speed at eighteen knots in a fair wind, had already been tested on the California run. In the Australian trade, they hit their stride, and in the first four years of the rush, one hundred fifty new clippers were built to meet the demand.

Meanwhile, there were problems at the diggings. Most of them rose from conflict between the free-spirited miners and the government agents. The agents had to enforce the licensing program, which called for a monthly fee of thirty shillings from each miner. The miners did all they could to avoid paying the fee and elude the "digger hunters" who came to collect it.

Gold was first discovered in New South Wales, but discoveries in Victoria quickly followed. Today, most Australian gold is mined in Western Australia.

The miners banded together and sometimes did battle with agents, even with government troops. Out of the bloodshed rose a cooperative spirit among miners that led to the formation of trade unions. Today unions are the leading political force in both Australia and New Zealand.

Meanwhile, as gold diggings in New South Wales grew, the colonial government in neighboring Victoria faced its own population problem. It seemed that everyone, including established farmers and tradesmen, had left Victoria to seek gold in New South Wales.

What Victoria neded was a rush of its own. In storybook fashion, the government offered rewards for anyone finding gold in the Victoria District. And in storybook fashion, gold was promptly discovered. Rich finds at Golden Gully, Bendigo, and Ballarat were well advertised. The focus of the Australian rush changed dramatically to Victoria.

In one two-week period, forty-five ships carrying a total of fifteen thousand gold-seekers left London bound for Victoria. And in less than a year, the population there had grown by one hundred thousand.

In the ten years following discovery, the Victoria fields yielded two hundred sixty-four million dollars worth of gold, and the diggings continued producing long after California gold was played out. Although production of Australian gold took a sharp decline soon after 1900, isolated discoveries continued. Gold is now being found in western Australia, and while it does not spark major gold rushes, it keeps hopes alive for new prosperity down under.

The Australian gold rush was just one of many that beckoned miners away from California. By one estimate there were as many as thirty rushes during the ten years that followed James Marshall's discovery. Most of them were false alarms, but reports of gold on British Columbia's Fraser River proved true. Some six percent of California's mining population made the long journey north. While some found enough gold to pay for their troubles, most soon discovered that the Fraser was no match for California's rivers.

Soon after, reports of gold on Pike's Peak started one hundred thousand hopeful gold-seekers toward the heart of the Rocky Mountains. Less than half completed the trip, and they found that the stories of a solid core of gold within the mountain were fiction.

The ores found at Pike's Peak were not rich and hardly plentiful enough to pay the wages of one hundred miners. Most other rushes of the era had the same, unhappy pattern—big promotion followed by small yields. But the discovery on western Nevada's Mount Davidson was different.

Stories about the discovery of the famous Comstock Lode differ. By one account, two brothers who failed to find gold in California, wandered into Nevada and found a rich vein. Both died, however, before they could register their claim.

A different account credits the discovery to two Irishmen—McLaughlin and O'Riley. They found black earth around a ground squirrel's hole, had the earth tested and found it was high grade electrum. McLaughlin reportedly sold out for thirty-five hundred dollars. O'Riley held on long enough to get forty thousand dollars for his interest.

Still another account credits Henry T. P. Comstock, a sheepherder and gold prospector known as "Old Pancake," as the force behind the discovery. Comstock and others laid claim to several sections of Mount Davidson, but they sold out cheaply in 1859 when they failed to find gold. They should have been looking for silver, for the Comstock Lode, as it became known, became the world's chief supplier of silver for the next twenty years.

But by 1890 a new gold rush was in the making.

Ounce for ounce, the gold of the Klondike caused more suffering for those who sought it than any other gold discovered in the nineteenth century. In fact, not since the days of the ancient slave mines have men gone through more physical torture in the quest for gold.

To get to the remote, upper reaches of Canada's Yukon River, miners risked starvation, frostbite, and exhaustion. Often, those who made it to the diggings were too spent physically to search for gold.

The rush was touched off by the discovery of a nugget about the size of a thumb on the banks of the Klondike, a tributary of the Yukon about two hundred fifty miles east of the Alaskan border. The man who found the nugget was a colorful prospector whose full name was George Washington Carmack but who was known along the Yukon as "Siwash George."

Born in a covered wagon that was crossing the plains, Carmack was the son of a Forty-niner who failed to find a fortune in California. Carmack went north to prospect in 1855 and decided to stay. He married the daughter of an Indian chief and took on the ways of the tribe, spending more time hunting moose and catching salmon than searching for gold.

Although other white men in the area sometimes derisively called him the "squaw man," Carmack did not go completely native. He had an organ in his cabin and he could play it well. His favorite magazine was the *Scientific American*, which was beyond the understanding of most other whites in the area.

One day while out on a fishing trip, Carmack and his two brothers-in-law, Skookum Jim and Tagish Charley, met Robert Henderson who had an unfortunate prejudice against Indians.

Henderson, who had searched for gold in Australia and the Rocky Mountains, told Carmack of a promising creek that might have gold because it drained off the Dome, highest mountain in the area. Henderson, not having time to investigate, said he would go shares for his tip if Carmack found any gold there.

"But I don't want any Siwashes (Indians) staking on that creek," Henderson reportedly said.

A few days later, Carmack investigated Rabbit Creek as Henderson suggested. And there, on August 17, 1896, the nugget that started the last great gold rush of the century was discovered. Carmack and his brothers-in-law staked claims and told Henderson nothing of the discovery.

News soon traveled up and down the Yukon that a rich strike had been made. Prospectors of the area, of course, were the first on the scene, staking claims in the choicest spots. The Rabbit was renamed Bonanza Creek, and the cluster of cabins that sprang up near the creek was called Dawson City.

While the first arrivals staked claims along the Bonanza, the latecomers investigated the smaller streams that flowed off the Dome. Five men, to become known as the "Bonanza Pups," made their fortunes on one of these creeks. One of the men, a barber, took fifty thousand dollars from his claim each summer for five years. Several other miners did exceptionally well that first season with claims that produced an average of eight hundred and fifty dollars a day. But then came winter.

Most Klondike gold was in river gravels that were

buried beneath a frozen overlay of mud and silt, five to thirty-five feet deep. After a day of digging, the miner usually built a fire in the bottom of his pit. Next day he would shovel out what thawed and build another fire.

On cold days, when the temperature dropped far below zero, just a few inches could be dug. The miners smeared a mixture of bacon grease and ash on their faces as a poor shield against the cold. When the temperature dipped to eighty below, picks and shovels would break if struck on a rock. Miners stayed in their thin-walled shacks burning tables, chairs, even vital sluice-box planks in the effort to keep warm. Rum offered some relief, but even it would freeze if the miner failed to keep the bottle warm.

The first year of the rush was mostly a local event, drawing prospectors who were already living in the Yukon or Alaska. Winter closed in before news of the discovery could draw miners from the outside world.

In the spring of 1897, however, two ships headed south with three tons of Klondike gold and several frostbitten but wealthy miners. One boat landed in Seattle, the other in San Francisco. From both ports the cry of "Gold!" spread across the land. The great rush began.

There were two routes to the diggings. One took you by ship to the Bering Sea and then by river steamer up the Yukon to Dawson. The second took you by ship to either Juneau or Skagway on the southern coast of Alaska and then by trail over the snowy mountains to the shores of Lake Bennett. The lake had to be crossed by raft and beyond the lake, White Horse Rapids had to be negotiated before you reached Dawson.

The first route was expensive, and if the Bering Sea were blocked by ice, it could take as long as eight months. The second route was cheaper but risky.

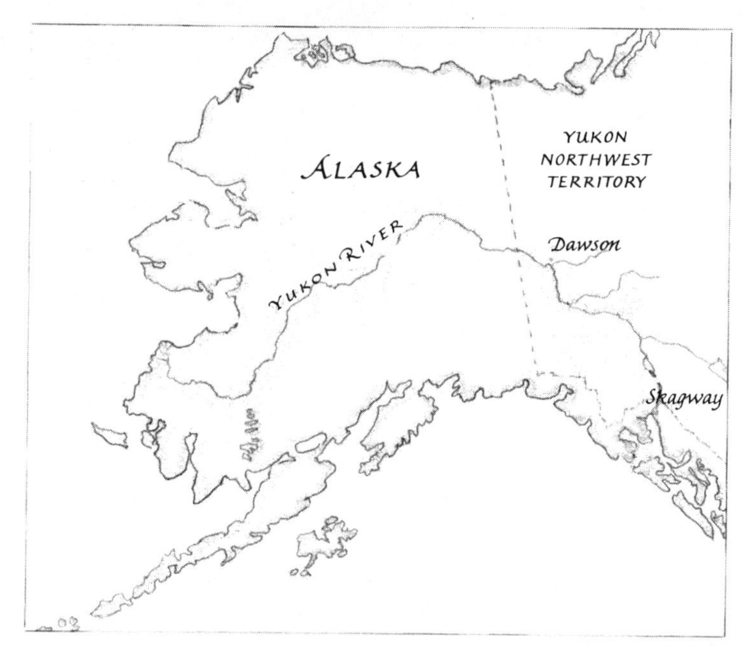

A miner could either reach Dawson by boat up the Yukon River or by packtrail from Skagway across the high passes to the upper reaches of the Yukon.

Most miners, in their innocence, chose the second route—six hundred miles overland to Dawson. Experienced Arctic travelers could make the journey safely in winter by dog sled. The miners hit the trail at the height of the spring thaw with horses as pack animals.

In the coastal towns, profit-hungry outfitters provided horses that should have been sent to the glue factory. Overloaded and hard driven by the inexperienced miners, the horses died by the hundreds. In the summer of 1897, just twelve of three thousand horses that left Skagway survived. The route over the mountain passes became known as Dead Horse Trail. It was lined with rotting carcasses of the poor animals.

Miners who lost their horses either turned back and bought new stock or else shifted as much of the animals' loads as possible to their own backs and pressed on. Ice, blizzards, temperatures at forty degrees below zero, hunger, and frostbite turned the journey into a nightmare. Many turned back. Some died. Of the one hundred thousand who tried to reach Dawson that first season, no more than forty thousand got through, and half of them were too weak to do any prospecting.

Dawson itself was no haven. Tents and thin-walled shacks were poor shelter in the Arctic winter. And new arrivals kept coming in all through the winter months. By the spring of 1898, supplies were desperately short. The first river steamer after the spring thaw rescued the new town from starvation. More boats followed.

In no time at all there were two banks, two newspapers, five churches, and scores of saloons, dance halls, and gambling parlors. In 1900, when Dawson's gold production peaked at forty-five thousand pounds, the town had steam heat, electric lights, a telegraph, and even a railroad. But by then the rush was over. The big companies with their huge dredges had tied up all the good claims, leaving no room for the hardy sourdough.

The Klondike experience and spirit, however, lives on in the stories of Jack London and verse of Robert Service, both of whom took part in the last great gold rush of the century. In describing the cruel loss of the horses, London wrote: ". . . men shot them and worked them to death and when they were gone went back to the beach and bought more. Some did not bother to shoot them, stripping the saddles off and the shoes and leaving them where they fell. Their hearts turned to stone—those which did not break—and they became beasts, the men of the Dead Horse Trail."

Service used humor to take the edge off the horrors of the northern winter. *The Cremation of Sam McGee,* perhaps his most famous poem, reads in part:

There are strange things done in the midnight sun
By the men who moil for gold;
The Arctic trails have their secret tales
That would make your blood run cold;
The Northern Lights have seen queer sights,
But the queerest they ever did see
Was that night on the marge of Lake Lebarge
I cremated Sam McGee.

The poem ends with the thawing McGee who asks that the door to his crematory oven be closed. He was just beginning to feel warm at last when the door was opened to create a draft.

Chapter Twelve

Gold Leaders

Today's leading gold producers are South Africa and Russia. The South African mines yield more than eight hundred tons of gold a year, about seventy-five percent of the western world's annual gold production.

We do not know exact production figures for Russia, but the output is probably somewhere between one hundred and three hundred tons a year. Deposits of Russian gold have not yet been fully exploited. In some remote regions, prospecting has just begun.

The oldest Russian sources are in the river valleys that drain into the Black Sea. The alluvial deposits from these rivers supplied gold all through the Persian, Greek, and Byzantine eras. And soon after yields from this region dwindled, a new source was found on the east slope of the Ural Mountains, the range that separates European Russia from Siberia.

The Russian tsars granted royal mining licenses that permitted members of the aristocracy to develop hard rock mines in the region. The aristocratic owners, already rich, grew even richer from gold. The mining was done by forced labor under harsh conditions. Although production continued many years, yields were modest compared with later Russian discoveries.

Early in the nineteenth century, a new, much richer resource was discovered further east in the Altai Mountains not far from Lake Baikal. Again, forced labor kept mining costs low, and again, a few Russian lords lucky enough to hold royal licenses grew fabulously rich.

While serfs hacked at frozen ground daily from 5 A.M. to 8 P.M., the lords of the gold mines feasted on imported caviar and oranges, sipped champagne, smoked Cuban cigars.

From 1814 to 1839, the Altai mines produced one hundred seventy-five thousand pounds of gold, and all through the 1840s, production never fell below seventeen thousand pounds a year.

Although the Altai still yields some gold, the focus today has shifted to the rivers of northern Siberia, rivers that flow into the icy Arctic Ocean.

Under the Communist regime, all gold, whether mined and smelted or still in the ground, belongs to the Russian government. And government spokesmen claim that this law does not have to be enforced. The people respect it.

It is a fact, however, that no prospector is allowed to search for gold alone. He must have others with him at all times. It is also a fact that two fist-size nuggets on display in a Siberian museum are facsimiles. When a western reporter recently asked why the real gold was not displayed, he was told by a museum official that it would be promptly stolen.

The government regularly sends teams made up of

twenty to one hundred prospectors into the field to search for gold. Their pay depends on how much gold they find. A successful prospector can earn as much as twelve thousand rubles or about fifteen thousand dollars a year.

Although hard rock mining has recently begun in the Altai Mountains, most of the Russian gold is found in river sands and gravels. Occasionally, a large nugget is recovered, but most gold is in grain or dust form.

South African gold is quite different. About two billion years ago a layer of gold-bearing sands and gravels some fifty miles long was covered by earth and rock. The weight of the overlying material eventually solidified the sands and gravel forming a kind of rock that geologists describe as a conglomerate.

The conglomerate layer or "reef" varies in thickness from a fraction of an inch to one hundred feet. Most of it still lies buried, but in some spots arches in the reef bring it to the surface. It was at one of these spots on a February day in 1886 that George Harrison and George Walker began quarrying rock for a house foundation. They were working on a farm belonging to the widow Anna Oosthuizen in a district known as Witwatersrand.

After the two men had broken a few rocks from the mass of conglomerate, they decided it was too crumbly to make good foundation material, but Walker, who had done some gold mining in Australia, decided to take a few samples home.

There he crushed the rock with a hammer, panned it in his frying pan, and found gold.

Harrison and Walker immediately staked a mining claim on the widow's property. Then Walker wrote to government officials in Pretoria, informing them of the discovery.

"I have a long experience as an Australian gold digger and I think it a payable field."

News of the discovery brought an eager army of prospectors to Witwatersrand which soon became known simply as the Rand. A cluster of miners' tents and shacks soon bore the name Johannesburg. Fifteen years later the city had a population of one hundred thousand gold seekers.

But the individual prospector did not fare well in South Africa. Most of the gold-bearing reef was buried deep in the earth. The rock, not rich in gold, had to be crushed and processed. In most cases, a ton of rock had to be mined and crushed just to get a quarter ounce of gold.

Large companies were needed to finance such operations. One firm spent eight million dollars just to open a mine. Clearly, South Africa was not the place for a lone prospector with limited finances.

The country already had well-established firms operating the diamond mines along the Vaal River in Kimberly. These firms had the capital and the foresight to exploit the newly discovered gold. They moved into the Rand and bought up all claims.

Harrison was one of the first to sell his share of the claim. He let it go for fifty dollars. A few years later he was killed by a lion. Walker, who sold his share for fifteen hundred dollars, died in poverty in 1924. By then, all South African gold production was in the hands of seven prosperous companies. Most of the labor was done by Bantu tribesmen who were imported to the region from fifty neighboring tribes. Low wages helped the companies hold profits high.

For a time, however, production was restricted by inadequate refining methods. The main problem was recovering gold that had combined with compounds of sulphur. Sulphuret gold, as it was called, could not be extracted with the old mercury process, and as the mines were sunk deeper beneath the Rand, sulphuret gold increased alarmingly. By 1890, it seemed that the South African gold industry would die young.

But a new process, patented by Scot inventors, John S. MacArthur and William Forrest, came to the rescue. Their smelting method, using cyanide compounds, recovered ninety-five percent of the gold mined. As the companies converted to cyanide smelting plants, production took a steady increase. In 1892, the Rand produced one million ounces of gold. In 1898, production hit four million ounces.

The Boer War interrupted mining, but in 1902, it resumed on a massive scale. Today, South Africa's annual gold output is measured in tons. Recent years have seen the figure soar over eight hundred tons of gold.

Although new government regulations have raised the salaries and improved living conditions for the native workers, profits for the companies remain high. More regulations, however, may be on the horizon. The unfairness of South African racial policies is all too clear at the mines.

A black worker is paid a hundred dollars a month plus food and lodging, but he must live in a compound where women and families are excluded. A white mine worker, usually a supervisor or an engineer, is paid about a thousand dollars a month and has an eight-room house provided at low rent for himself and his family.

The white man's work week is limited to forty-eight

hours. The black man is expected to work sixty hours in five, twelve-hour shifts each week.

In recent years, neighboring tribes have refused to supply new recruits for the mines. The big companies may soon be forced to provide more equitable pay and better conditions for the black miners.

Even so, it is likely that South African gold mining will remain profitable for many years to come.

Chapter Thirteen

Golden Freedom

Most of the gold dug from the mines of South Africa goes to the big markets in Zurich, Switzerland, and London, England. There some of it is converted into gold bars no bigger than match boxes.

These small bars have a special market. They are flown east to a bustling city on the west shore of the Arabian Gulf. Dubai used to be nothing but a little village that a few struggling pearl fishermen called home. Today it is a large commercial center boasting twenty banks.

Part of the prosperity comes from oil, but Dubai owes its commercial beginnings to those small gold bars and to the governments of Iran, Pakistan, and India. Laws in these countries restrict importation and limit possession of gold.

But because gold is valued almost as much as life itself in these countries, the laws are impossible to enforce. In India, there are five hundred million Hindus with a five thousand-year-old tradition that equates gold with virtue and purity.

Small gold bars, no bigger than a matchbox, are used in the smuggling trade between Dubai and India.

When a Hindu dies, a piece of gold must be placed in the mouth. When a Hindu child is born, the father must touch it with gold. A Hindu gives gold to get rid of sins and wears gold for prosperity and luck. A Hindu doctor prescribes gold concoctions to cure just about every ill.

When a girl marries, she wears her dowry in gold jewelry. The marriage would not be possible without at least two bracelets, a set of earrings, a necklace, a forehead ornament, a tiara, a chain running from a nose ring to one ear, and a *Ratanchoor* on each hand that links with golden chains the rings on every finger to golden wristbands.

In some regions of India, the gold must be pure, twenty-four karat. Brides put most of their jewelry in safe-keeping after the wedding. The gold serves as the family insurance policy. And in a country where the government has been known to devalue currency or seize private bank accounts overnight, hoarding gold makes good sense. A person with gold is free in a very real sense from government controls and sudden actions.

Officials in Iran, Pakistan, and India say that hoarding gold takes currency out of circulation and retards the economy. But the laws restricting gold are useless.

They do, however, help the merchants of Dubai. A fleet of boats, manned by expert seamen, leaves the port regularly. Each boat carries a cargo of the small bars. As the ships sail east, seamen busy themselves with needle and thread. They sew the bars carefully into the lining of canvas jackets.

The Dubai merchants do not regard themselves as smugglers. And technically, they are not. The jackets, bundled into bags, are passed to eager hands on other boats well outside the sea boundary of any country. It is the chore of those on the other boats to get the illegal gold into their countries.

Back at Dubai, suitcases of currency, often United States dollars, which were given in exchange for the gold, are counted and banked. More small bars are ordered. Plans for a new voyage are made.

Despite soaring gold prices in recent years, the Dubai trade continues to flourish. It is just one region of the world where gold is being smuggled or being sold under the counter.

In Colombia, high in the Andes of South America, there is an unusual gold trade in Indian artifacts. The government pays well for such treasures and professional grave diggers who used to sell artifacts illegally to private collectors now sell most of them to the government. As a consequence, the Gold Museum in the capital of Bogotá is a showcase of native gold and the pride of Colombia. Some artifacts, however, are still offered privately. A few of these might be genuine, but most are skillful imitations of ancient figurines and ornaments.

A visitor to Bogotá who stands on the right street corner or says the correct words to a taxi driver will soon be offered a choice of golden "relics." The seller will claim they are real antiquities, but the chances of them being fake are high.

Large dredges are still taking gold from Colombian rivers, and some of it finds its way into the counterfeit relic trade. Natives also mine gold with primitive panning and sluicing techniques in Colombia's remotest rivers, but little of this gold reaches Bogotá. It is smuggled to Ecuador where it now fetches a higher price than the Colombian banks offer.

In Egypt, where gold was once held by the pharaohs only, today's peasants demand gold. They want to hide it away where it cannot be traced by the government. It will not rust or mold. It will always be there.

Each year after the harvest is in, dealers in Cairo keep busy selling their gold to farm workers. Many of the buyers plan one day to give the gold to their grandchildren. It will be a valuable gift.

Iraq bans private importation of gold, but the natives there demand it, and thanks to a brisk smuggling trade, they get it. Gold in the form of counterfeit fifteen-ruble pieces is even smuggled into Russia where it is purchased by hoarders.

What keeps gold moving through both legal and illegal channels? It all comes down to one simple explanation. People want gold.

Governments, concerned about economies of their countries, have repeatedly tried to change attitudes toward gold and control its use. There are logical reasons for this. Hoarding takes money out of circulation, and hoarded gold earns no interest, pays no taxes.

Bankers of the sixteenth century guarded their wealth just as zealously as they do today. When gold was in short supply, loans were difficult to obtain,—at least that's the way this period drawing makes it seem.

In most cases, regulations fail, especially when they run counter to traditions. When it comes to gold, tradition is stronger than rationality.

For hundreds of years, any European lucky enough to obtain gold could take it to a licensed minter and have it converted to coins. The minter kept a small coinage fee, and the government took a fee for use of the dies that gave the coins their recognizable value.

The system had its defects. For instance, there were so many different mints producing such a wide variety of coins that it was impossible to keep track of, let alone control, the money in circulation. In Germany alone, during the 1300s, there were more than six hundred mints, each making distinctive coins. As a consequence of the profusion of minting throughout the Old World, only a few coins such as the florin, the ducat, and the daric gained international recognition.

Another defect was that a shortage of gold, the most common condition, directly restricted the money supply. Thus, when there was not enough gold to keep the mints producing at a normal rate, the economy faltered. The direct link between gold supply and the economy could and often did lead to serious consequences.

Yet the licensed minting system, despite its defects, lasted well into the 1600s and left a permanent imprint on monetary customs.

To be accepted, the conversion to paper money thus had to be a gradual one. But silver or gold certificates circulating in place of coins were certainly more convenient. And as long as a government stood ready to redeem them for their face value in the precious metal, the certificates could be trusted.

But it took the gold discoveries of the last century in California, Australia, the Yukon, and South Africa to give governments of the world enough gold to support a sound paper money system. The United States and India long maintained a bimetal system, issuing certificates for both silver and gold. By the beginning of this century, however, all countries were on a gold standard.

It was a good thing, too good to last.

With the world on a gold standard there was practically no speculation in foreign currency. Importing and exporting of gold could be carried out almost anywhere without restriction. The gold standard thus brought stability in both domestic and foreign trade.

Countries that exported more goods than they imported enjoyed a growing economy and an ever-accumulating gold supply. Our own Fort Knox is testimony to our recent history of a favorable balance of trade. Countries that imported more than they exported paid the price in gold. But because gold grew scarce in these countries, prices dropped as importers competed for trade.

The gold standard, in effect, fixed the price of gold. For a hundred years, from 1834 to 1934, the United States fixed the price at $20.67 an ounce.

World War I killed the gold standard. The war forced governments to overspend on weapons. England went off the gold standard in 1919, and began printing more notes than it could redeem in gold. The United States clung to the gold standard until 1933, when the post-war depression brought on huge government spending programs.

Politicians, who know they can win votes by spending tax dollars in their districts, were happy to see the gold standard dead and buried. With the standard out of the way, the government could simply print more money to cover ever-expanding spending programs. The politicians got the credit and the votes.

In 1925, England, realizing it had made a mistake, tried to return to the gold standard. But it was too late. Spending had exceeded the gold supply by such a huge margin that the country had no hope of backing the paper money then in circulating with anything more than a promise.

Meanwhile, the United States, along with most other countries, adopted an exchange standard and made private possession of gold in any form other than jewelry and dental fillings illegal. The ban on private ownership came about during the depression and was designed to keep wealth from being frozen in gold hoards.

The exchange standard allowed banks to use key currencies, usually dollars, to buy reserves of gold that served to back only in part the paper money in circulation. Meanwhile, most governments continued over-spending and printing more money to pay their bills. The practice was bound to lead to trouble.

Fortunately for the United States, however, the other major commercial countries ran into trouble first. World War II hastened the financial collapse of Europe. Emergency measures and an overhaul of the monetary system were needed at once.

The United States adopted the Marshall Plan to put Europe's economy on its feet. Meanwhile an International Monetary Conference was called to Bretton Woods, New Hampshire. In 1944, more than one thousand conference delegates adopted a new system. Gold from now on would be used only to pay debts between governments. No private citizens anywhere would be allowed to redeem gold with paper money.

In practice, the United States had been following this system for several years. The conference, however, made it a worldwide system, and thereby established the United States dollar as the leading international currency.

In dealing with other countries, we backed our currency at the rate of thirty-five dollars an ounce of gold, and with a healthy post-war balance of trade, our gold reserves soared at an unheard of rate. As farm produce and trade goods flowed out, the gold flowed in. By 1949, just five years after Bretton Woods, our gold reserves peaked at 24.7 billion dollars.

For twenty-five years, the United States dollar was "good as gold." But our economic power could last only as long as we had enough gold to back all the dollars circulating overseas.

The Army, tourists, foreign aid, and foreign business investments helped hasten the end. By 1958, the free spending of United States dollars abroad made it clear:

the international gold standard, like the earlier domestic standard, was dead. It took ten years, however, for international bankers to put it to rest, and even then, it was with stopgap action.

The bankers set two prices for gold—thirty-five dollars for gold used to trade paper money and a fluctuating price for gold offered in the open market. This action was followed almost immediately by Britain's decision to devalue the pound by 14.3 percent. Investors, fearing another devaluation, began buying the one thing that was certain to have lasting value—gold.

In December, 1971, President Richard Nixon ordered devaluation of the dollar by eight percent. The move followed a year that showed the first United States trade deficit in seventy-eight years. We were importing more goods than we were exporting. Gold reserves had dropped to ten billion dollars. Our political leaders, still spending more than they could raise in taxes, had built the government's annual deficit to a record 23.2 billion dollars

Countries that had long relied on United States dollars began cashing them in for gold. In 1973, a further dollar devaluation of ten percent launched a modern-day gold rush.

In 1974, our government made it legal again for United States citizens to invest in gold. Contrary to expectations, the gold market was not buried in purchase orders.

Americans, not used to buying gold, entered the field gradually. But as the price of gold rose steadily on the free market its attraction strengthened. Speculators could make thousands of dollars on the rising gold market in just a few days of investing. More conservative investors used gold as a hedge against inflation. After three years of fluctuation, gold was selling for $177 an ounce and rising.

By tradition, the London banking offices of N. M. Rothschild & Sons, Ltd., sets the price of gold daily. The price is followed closely by gold trading firms in Zurich, Paris, Frankfort, Milan, Beirut, Bombay, Hong Kong, and New York. Transactions are made by bars of gold, which range in weight from three hundred sixty to four hundred thirty ounces. The bars usually remain in storage deep in banking vaults. The changes in ownership are simply recorded on bank ledgers.

In the late 1970s, the price of gold soared above six hundred dollars an ounce. The peak came during high political unrest in the world. Unrest continued into the 1980s, but by 1982, the price of gold was fluctuating between four hundred and five hundred dollars.

The high price for gold has stimulated both mining and treasure hunting. In many cases gold deposits that were once too lean to be mined feasibly are now being worked.

In the Battle Mountain region of central Nevada, for instance, the Duval Corporation is bringing both long known and recently discovered gold deposits into production at a cost of two hundred dollars per ounce. A few years ago, such a venture would have been unthinkable.

Treasure hunters have revived old dreams and begun searching for many legendary hoards ranging from Nazi gold thought to be hidden at the bottom of Alpine lakes to gold that Alexander was said to have buried in various places during his rapid conquest of the Near East.

The largest gold recovery from a sunken ship was accomplished north of the Arctic Circle in 1981, a year when diving technology and high gold prices combined to make the venture possible.

The gold, five tons of it worth eighty million dollars, went down with the cruiser *HMS Edinburgh*, which was sunk by two German torpedoes in 1942. The ship lay under eight hundred feet of frigid water, virtually unreachable for almost forty years. Then a British salvage company, after two years of search and an investment of four million dollars, located the wreck and recovered the gold.

Divers, equipped with suits warmed by hot water lines, worked from a diving bell hung forty feet above the wreck. Water pressure at the *Edinburgh* was three hundred fifty pounds per square inch, but the men were able just the same to work their way through a storeroom of bombs and shells to locate 431 of the original 465 gold bars.

Because the gold had been sent from Russia to Britain as payment for weapons and because both governments had insured the gold against loss, it was government property. The salvage firm, however, was paid thirty-six million dollars for its efforts before Britain and Russia divided the balance of the treasure. It was a good return for a four million-dollar investment.

Another kind of gold investment, although not so spectacular, is nonetheless attracting a great deal of attention. And indeed, some people have already made fortunes dealing in gold coins.

The corona of Austria, the peso of Mexico, the maple leaf of Canada, and the krugerrand of South Africa are now being purchased as fast as they can be minted. A coin market, which began operating internationally in the early 1980s, gives hourly quotations on the prices of all gold coins.

Strangely, this phenomenon arose at a time when most governments were trying to remove their monetary systems as far as possible from any relationship with gold. Bankers have even developed a substitute for gold.

The International Monetary Fund, formed at the

Bretton Woods Conference back in 1944, recently established a unit called Special Drawing Right or SDR to replace gold. The international value of an SDR is readjusted by computer daily in relation to major world currencies. For use only as an exchange unit between governments, SDRs have gradually gained acceptance. Unlike gold, SDRs can earn interest. And they don't have to be weighed, stored in a vault or guarded. But how will SDRs hold up in a world crisis? No one can make any predictions. But you can always be sure of gold.

Gold will hold its value. Despite the efforts of major governments to play down its importance, it will continue to be the safest "money" you can own. And this will remain true as long as there are people in this world like you and me, normal people who desire gold.

SELECTED BIBLIOGRAPHY

ALLEN, GINA. *Gold!* New York: Thomas Y. Crowell Company, 1964.

ARNOLD, OREN. *Marvels of the U. S. Mint.* New York: Abelard Schuman, 1972.

BEEBE, LUCIUS, AND CLEGG, CHARLES. *San Francisco's Golden Era.* Berkeley, CA: Howell-North Books, 1960.

BOXER, C. R. *The Golden Age of Brazil, 1695–1750.* Berkeley, CA: University of California Press, 1962.

DURANT, WILL. *Our Oriental Heritage.* New York: Simon & Schuster, 1954.

———, *The Life of Greece.* New York: Simon & Schuster, 1939.

EINZIG, PAUL. *Primitive Money in Its Ethnological, Historical, and Economic Aspects.* New York: Paragon Press, 1966.

GREEN, TIMOTHY. *The World of Gold Today.* New York: Walker & Company, 1973.

HAHN, EMILY. *Love of Gold.* New York: Lippincott & Crowell, 1980.

KERBY, ELIZABETH POE. *The Conquistadors.* New York: G. P. Putnam's Sons, 1969.

LYTTLE, RICHARD B. *People of the Dawn.* New York: Atheneum, 1980.

———, *Waves Across the Past.* New York: Atheneum, 1981.

MARX, JENIFER. *The Magic of Gold.* New York: Doubleday & Company, 1978.

MASON, BRIAN. *Treasures Underground.* New York: Home Library Press, 1960.

TAPER, BERNARD, ED. *Mark Twain's San Francisco.* New York: McGraw-Hill, 1963.

VILAR, PIERRE. Translated by Judith White. *A History of Gold and Money.* Atlantic Highlands, N.J.: Humanities Press, 1976.

WADE, WILLIAM W. *From Barter to Banking: The Story of Money.* New York: Crowell and Collier Press, 1967.

WHITE, PETER T. "The Eternal Treasure—Gold." Washington, D.C., *National Geographic,* January, 1974.

INDEX